HOW CAN YOU TELL IF A SPIDER IS DEAD? AND MORE *Moments of Science*

HOW CAN YOU TELL IF A SPIDER IS DEAD? AND MOR

EDITED BY *Don Glass*

Moments of Science

WITH CONTRIBUTIONS BY *Stephen Fentress, Barbara Bolz, Eric Sonstroem, and Don Ulin*

INDIANA UNIVERSITY PRESS • BLOOMINGTON AND INDIANAPOLIS

The paper used in this publication meets the minimum
requirements of American National Standard for
Information Sciences—Permanence of Paper for Printed
Library Materials, ANSI Z39.48-1984.

Manufactured in the United States of America

Library of Congress Cataloging-in-Publication Data

How can you tell if a spider is dead? : and   more
    moments of science / edited by Don Glass.
        p.   cm.
    "This book is based on the radio series 'A Moment
of Science'"—Pref.
    Includes index.
    Summary: Essays explore the world of science,
discussing such topics as the shape of the Earth, the
speed of tornado winds, and the iron content of
human milk.
    ISBN 0-253-33068-8 (cl. : alk. paper).— ISBN 0-
253-21020-8 (pbk. : alk. paper)
    1. Science—Popular works. 2. Science—Miscellanea—
Juvenile literature. 3. Moment of science (radio program)
[1. Science—Miscellanea.] I. Glass, Don, date. II.
Moment of science (Radio program)
Q162.H792   1996
500—dc20   95-50598

    2   3   4   5   01   00   99   98   97

To Paul Singh, whose inspiration and entrepreneurial spirit got it all started.

Contents

x

Preface

Boyce Rensberger, in his book *How the World Works*, states that we are all born scientists. And it's true—as infants and young children we are intensely curious about the world around us. We want to explore every nook and cranny. We test gravity by throwing our food on the floor, or defy it by throwing the food at someone. For many this intense curiosity wanes with age. For others the curiosity remains, but other interests prevent pursuing it. For a few the curiosity of childhood develops into a professional career as a scientist.

This book is based on the radio series *A Moment of Science*®, and both are designed for the young whose curiosity hasn't waned, and for those of us whose curiosity is still there but who have chosen other, non-scientific lives. We like to say that both the radio programs and the book are for those with unlimited curiosity but a limited amount of time. Each piece is a self-contained bit of enlightenment—you may read as few or as many as you want, and in any order.

Steve Fentress was the first and sole writer for *A Moment of Science* when it began in 1988, and his scripts constituted the first compilation, *Why You Can Never Get to the End of the Rainbow, and Other Moments of Science*. After he left Bloomington he continued to write scripts, many of which are included in this book. But the time pressures of his new career and life precluded his sustaining the series alone, so in the subsequent years other writers have worked with *A Moment of Science*. Those included here are Don Ulin, Barbara Bolz, and Eric Sonstroem, who have carried on the tradition Steve established by continuing to write scripts that are light, concise, entertaining, and informative. I hope that my contribution has been a positive one as well.

There are those who might not believe this, but science can be fun and captivating, and I hope this book will prove it.

Don Glass

Acknowledgments

This is the "would-not-have-been-possible-without" section, and even though the phrase might have been overused over the years, it is certainly true. Whether through financial, technical, or moral support, it took the combined efforts of many to make the radio series happen, and by extension this book (as well as its predecessor, *Why You Can Never Get to the End of the Rainbow, and Other Moments of Science*).

The material in this book came directly from the scripts for the radio series *A Moment of Science*®. The programs have been produced and aired nationally since 1988 by WFIU-FM at Indiana University in Bloomington.

During the period from which the scripts in this book were taken, much-appreciated corporate underwriting came from the Indiana Corporation for Science and Technology and Exmin Corporation. Gratitude is extended to Indiana University for the support provided by the Office of Research and the University Graduate School, the Office of the Chancellor in Bloomington, the Office of the Vice President, the Office of the President, and the Radio and Television Services.

Producing a science radio series or compiling a book on science necessitates that both live up to the implication that the information contained in them is correct, at least to the best of the knowledge at the time. Scientists who volunteered their time and expertise to help ensure accuracy for these pieces were Jeffrey Alberts, Moya Andrews, Susan Berg, Geoffrey Bingham, Jessica Bolker, William Boone, Jesse L. Byock, John Carini, Andrew Carleton, John Castellan, Ken Caulton, James Craig, Marti Crouch, Vladimir Derenchuk, Russell Dukes, David Easterling, Heather Eisthen, Andrew Ellington, John Ewing, Gerald Gastony, Steven Girvin, Michael Hamburger, George Hegeman, Margaret Intons-Peterson, Susan Jones, John Kissel, William Krejci, Ken Kuikstra, Alice Lindeman, Michael List, Charles Livingston, Alan Longroy, Bruce Martin, Anthony Mescher,

Lawrence Montgomery, Harold Ogren, Robert Peterson, Henry Prange, Rudolf Raff, Stephen Rennard, Edward Robertson, Unni Rowell, Bill Rowland, Steven Russo, Dolores Schroeder, Marc Schuckit, Sarita Soni, Catherine Souch, Joel Stager, Joseph Steinmetz, Milton Taylor, Esther Thelen, Larry Thibos, Christopher Tyler, Eugene Weinberg, Jeffrey White, David Williams, and David Wise.

Two people who had nothing to do with this book deserve recognition here because they get so little otherwise: Pat Hawkins Smith and John Shelton of WFIU's engineering and production staff. John has provided superb technical expertise, and engineered a number of programs. For years Pat has been the ever-present person on the other side of the glass (no pun intended), engineering the recordings, listening for technical glitches, as well as offering much-appreciated "esthetic" advice.

Special gratitude is extended to John Woodcock in the Indiana University English Department for his years of volunteer association with *A Moment of Science,* and the countless hours he has spent reviewing hundreds of scripts to help us stay on target at making science fun and understandable to everyone. Even when one of us thought we had gotten the text just right, John had an incredible knack for making suggestions that made it better.

Read on now and enjoy the fruits of our labors.

HOW CAN YOU TELL IF A SPIDER IS DEAD? AND MORE *Moments of Science*

The Shape of the Earth

Ancient civilizations described the earth as a flat plate, a dome, or even a huge drum. Today, with the help of photographs from space, we know the earth is a sphere. Yet even by the fourth century B.C., Greek astronomers with no more evidence than you could collect in your own backyard believed that the earth was spherical.

Aristotle, for example, based his argument for a spherical world on the shape of the earth's shadow during a lunar eclipse. A lunar eclipse occurs when the sun and the moon line up on opposite sides of the earth. When that happens, the shadow of the earth moves across and eventually covers the moon. Although the moon is too small to show us the entire shadow of the earth, the edge of the shadow that we can see is always curved, regardless of where the moon and sun are in relation to the surface of the earth. The only shape that could cast a round shadow from every angle is a sphere. A round, flat plate would cast a round shadow only if the light were coming from above or below; from any other angle, the shadow would be an oval or a straight line.

In fact, anyone driving west across the country will notice that the peaks of the Rocky Mountains appear before the foothills. If we say that this is because the foothills are still below the horizon, we are implying—quite correctly—that the curvature of the earth gets in the way. At closer range, there is no horizon between us and what we are looking at, and so the earth appears flat. But if the earth were really flat, the only limit to how far we see would be the power of our vision, and on a clear day the full height of the Rockies would be visible from Kansas.

Cohen, Morris R., and I. E. Drabkin. *A Source Book in Greek Science.* Cambridge: Harvard University Press, 1958.

Kuhn, Thomas S. *The Copernican Revolution.* Cambridge: Harvard University Press, 1957.

A Rising Fastball

Suppose a baseball pitcher releases the ball with a sharp downward snap of the wrist. That snap of the wrist imparts backspin to the ball. Backspin is opposite in direction to the spin of a ball just rolling on the ground. If the ball has backspin, then the ball's top surface moves in the same direction as the flow of air over the ball, and the bottom surface moves in the opposite direction to the flow of air. So, because of friction between the surface and the air, the top surface of a ball with backspin helps the air along, so to speak. And the bottom surface slows the air down.

Now, a famous principle: A fast-moving air stream has lower pressure than slow-moving or still air around it—a principle named after the eighteenth-century physicist Daniel Bernoulli, who discovered it.

From the point of view of a ball with backspin, the air on top moves faster and therefore has lower pressure than air on the bottom. The result is a force pushing the ball up. A baseball thrown hard enough, with enough backspin, will rise—or at least seem to be suspended—as it approaches the plate. That's your basic fastball.

This effect was first described in 1852 by the German physicist Heinrich Magnus. He was thinking about spinning cannonballs rather than spinning baseballs, but the principle is the same: Spin makes air flow faster along one side of the ball than the other. That makes a difference in air pressure, which in turn influences the flight path.

Gray, H. J., and Alan Isaacs, eds. *A New Dictionary of Physics.* London: Longman, 1975.
"Magnus Effect." In *Oxford English Dictionary,* 2nd ed. New York: Oxford University Press, 1989.
Prandtl, Ludwig. *Essentials of Fluid Dynamics.* London: Blackie, 1952.

Pine Tar Home Runs

It's against the rules of baseball to put pine tar on the fat part of the bat. Why? What's the unfair advantage of a tarred bat?

A couple of Tulane University physicists, Robert Watts and Steven Baroni, published an article about this in a journal for teachers. They pointed out that tar increases friction between bat and ball, so a tarred bat can give the ball faster spin. Specifically, an uppercut ball has backspin. Other things being equal, a fly ball with fast backspin will travel farther—dozens of feet farther, according to the physicists' calculations—than a fly ball with no spin.

The reason takes us back to Bernoulli's principle: A fast-moving air stream has lower pressure than still air around it. The top surface of a backspinning, traveling ball sweeps air back over the ball. From the ball's point of view, the air on the top moves faster and therefore has lower pressure than the air on the bottom. The result is a force pushing the ball up. A ball with fast backspin seems to float through the air.

Watts and Baroni concluded that the best strategy for hitting home runs is to hit the ball slightly below center, so as to give the ball a fast backspin rate and a low launch angle. A tarred bat is illegal but gives faster spin.

Watts, Robert G., and Steven Baroni. "Baseball-Bat Collisions and the Resulting Trajectories of Spinning Balls." *American Journal of Physics,* January 1989.

The Force of a Tornado

Tornadoes are known for their powerful winds, but the winds alone cause only part of the damage. In addition, the high winds create an area of extremely low pressure that can actually cause buildings to explode.

To understand how the winds cause the pressure to drop, we can review Bernoulli's principle, the same principle that explains the curve ball. Bernoulli's principle states that a fast-moving air stream has lower pressure than slow-moving or still air around it.

So when the very high winds of a tornado blow over a house, they cause the pressure to drop to a level far below that of the still air inside the house. The higher pressure in the house can then force the roof off the top of the house. As the roof comes up, the tornado winds pick it up and carry it away.

To see how this works on a much smaller scale, hold a dollar bill on the back of one hand. With the other hand, hold the edge that is away from you so the bill doesn't fall off. Now, blow hard across the top surface of the bill, and it should lift up off your hand. If you blow down onto the paper it won't work, but if you blow parallel to the surface of the bill, you should be able to decrease the air pressure above the paper enough to cause it to rise. Once it rises slightly, your breath will catch the underside of the bill and blow it back across your other hand.

When a window is broken in a tornado, the glass usually lands outside the house. This is because the force is exerted from the area of higher pressure inside the house toward the area of lower pressure outside the house.

Ahrens, C. Donald. *Meteorology Today: An Introduction to Weather, Climate, and the Environment.* St. Paul: West Publishing Company, 1991.

Anemia

Most cells in the body are capable of repairing themselves and of reproducing new cells just like themselves. One type of cell, however, for which this is not the case is the red blood cell. The red blood cell has no nucleus or chromosomes, and so is incapable of either reproducing or repairing itself. Instead, the body keeps up a constant supply of these very important cells in the bone marrow to replace red blood cells as they wear out.

Red blood cells are responsible for carrying oxygen from the lungs to the other cells, and for carrying carbon dioxide back to the lungs. Normally the body keeps an appropriate number of red blood cells, but in some people the bone marrow either produces too few or produces defective cells. This condition leads to a loss of energy as the body's cells cannot get enough oxygen. The name for this condition is anemia.

The most common cause of anemia is poor diet. Pernicious anemia is caused by a deficiency of vitamin B_{12}. Vitamin B_{12}, which is an essential component of red blood cells, is available only from animal products.

Iron is also an essential part of red blood cells. Iron deficiency is most common among children, whose pool of red blood cells is increasing. Because all the iron in red blood cells is recycled by the body when a cell is destroyed, adults need iron only if they have lost blood.

If the body does not have enough iron, red blood cells are still produced, but they are small and less effective.

Still other forms of anemia are hereditary and strike certain segments of the population more often than other segments. These forms include sickle-cell anemia, primarily among people of African descent, and thalassemia, found mainly among people of Mediterranean descent.

Newton, Tennis, and the Nature of Light

The seventeenth-century English physicist Isaac Newton allowed a beam of sunlight to pass through a triangular glass prism. A rainbow of colors emerged from the prism and fell on the wall opposite the window. Newton wondered why different colors fell on the wall in different places. He wrote: "I began to suspect, whether the rays, after their trajection through the prism, did not move in curve lines, and according to their more or less curvity tend to divers parts of the wall. I remembered that I had often seen a tennis ball, struck with an oblique racket, describe such a curve line."

A tennis ball struck obliquely, or a baseball thrown with a snap of the wrist, curves because it is spinning. As the ball moves through the air, spin makes air flow faster along one side of the ball than the other. Pressure is lower on the side where the air moves faster. The ball is deflected toward the low-pressure side.

Newton toyed with the idea that light might be made of little balls that were somehow made to spin while passing through the prism. Then these balls might follow curved paths from the prism to the wall. Newton soon rejected the curve-ball idea when he saw that colored light emerged from the prism in straight lines, not curves. Centuries later, physicists learned that particles of light, the so-called photons, bear almost no resemblance to little balls.

But Isaac Newton's attempt to relate light to tennis balls, even though it turned out to be wrong, shows how a first-class thinker searched through all his experience for a picture to help him understand a puzzling phenomenon.

5

Feynman, Richard P. *QED: The Strange Theory of Light and Matter.* Princeton, N.J.: Princeton University Press, 1985.

Newton, Isaac. "Dispersion of Light." In *Philosophical Transactions, Abridged*, vol. I
(1672). Reprinted in *Moments of Discovery*, vol. I, ed. George Schwartz and Philip
W. Bishop. New York: Basic Books, 1958.

Stereo in 1881

If you had attended the 1881 Electrical Exhibition at the Palace
of Industry in Paris, you might have stood in line for one of the
most popular demonstrations at the Exhibition—a telephone,
conveying music from the stage of the Paris Opera. When your
turn came, you would have placed a receiver over each ear and
listened to a few minutes of whatever was being performed at the
opera at that moment.

    This might have been your first chance to hear a telephone in
person, since in 1881 the telephone was a recent invention. But
even people familiar with telephones noticed something peculiar
about that demonstration at the Paris Electrical Exhibition. The
sound was reported to have "a special character of relief and
localization" that a single telephone receiver could not produce.

    The sound had this special character because of the way this
particular telephone system was hooked up. The receiver over the
listener's left ear was connected to a microphone on the left side
of the opera stage, and the receiver for the right ear was connected
to a microphone on the right side of the stage.

    A contemporary article in the magazine *Scientific American*
attempted to describe the resulting effect with such terms as
"binauricular auduition" [*sic*] and "auditive perspective." Today
we'd use the word "stereo," from a Greek word meaning solid, to
describe the sound.

    It was not until 1957 that the first stereo music recordings were
sold to the public. But back in 1881, if you were lucky, you might
have been one of the very first people to hear music transmitted in
stereo.

Clark, H. A. M.; G. F. Dutton; and P. B. Vanderlyn. "The 'Stereosonic' Recording and
Reproducing System: A Two-Channel System for Domestic Tape Records." *Journal
of the Audio Engineering Society*, April 1958. Reprinted from *Proceedings of the
Institution of Electrical Engineers*, September 1957.
"The Telephone at the Paris Opera." *Scientific American*, December 31, 1881.

6

A Water Magnifier

Punch a hole about an eighth of an inch in diameter in the bottom of a paper or foam-plastic cup. Now push the cup down into a deep bowl of water. Look down into the cup while you push it into the water. Of course you'll see water come in through the hole you punched. But as the cup fills, you'll notice something else. You'll notice that the hole in the bottom of the cup appears magnified. You can make the hole appear bigger by pushing down harder on the cup. Push down more gently, and the magnification is reduced.

This happens because rays of light are bent when they cross from one medium to another. In this case, rays of light that make up the image of the hole in the bottom of the cup are bent as they cross from water into air. The rays are bent in just the right way to create a magnified image of the hole from your point of view.

The reason those light rays are bent in that particular way has to do with the shape of the water surface. When you push down on the cup, water spurts upward through the hole in the bottom. That upward-spurting stream of water makes a bulge in the water surface more or less like the bulge in the surface of a glass magnifying lens. By varying the downward pressure on the cup, you can vary the size of the bulge in the water surface. That, in turn, varies the amount of magnification.

McMath, T. A. "Refraction—A Surface Effect." *The Physics Teacher,* March 1989.

Bowled Over by a Sound Wave

In the 1870s the Irish physicist John Tyndall was famous all over Great Britain as a lecturer on science. One of Tyndall's lecture demonstrations showed how sound travels through air. Tyndall described it like this:

"I have here five young assistants, A, B, C, D, and E, placed in a row, one behind the other, each boy's hands resting against the back of the boy in front of him. I suddenly push A; A pushes B, and retains his upright position; B pushes C; C pushes D; D pushes E. Each boy, after the transmission of the push, becoming himself erect. E, having nobody in front, is thrown forward. Had he been standing on the edge of a precipice, he would have fallen over; had he stood in contact with a window, he would have broken the glass; had he been close to a drum-head, he would have shaken the drum. We could thus transmit a push through a row of a hundred boys, each particular boy, however, only swaying to and fro.

"Thus, also, we send sound through the air, and shake the drum of a distant ear, while each particular particle of the air concerned in the transmission of the pulse makes only a small oscillation."

If you try John Tyndall's demonstration yourself, you'll see that each person in line can't help pushing the next person. It's a reflex to keep from being knocked over.

John Tyndall had a knack for memorable physics demonstrations. He believed they were important. He wrote: "Scientific education ought to teach us the invisible as well as the visible in nature; to picture with the eye of the mind those operations which entirely elude the eye of the body."

"Tyndall." In *Dictionary of Scientific Biography,* Charles Coulston Gillispie, Editor-in-Chief. New York: Scribner, 1971.

Tyndall, John. *Sound.* 1867. Reprinted in *The Science of Sound.* New York: Philosophical Library, 1964.

How Fast Are the Winds of a Tornado?

The speed of tornado winds is difficult to estimate for two reasons. For one, weather instruments are often destroyed in a tornado. The other is that the winds are traveling at different speeds in different parts of the tornado.

Tornado winds travel in a circle around the center but, like any weather system, the tornado itself is also moving. Because the winds of a tornado travel in a circle, the winds on opposite sides of the tornado are always traveling in opposite directions. So, if the wind at one point is traveling due south, the wind on the opposite side is traveling due north.

Imagine now that a tornado with 150-mile-per-hour winds is traveling due north at 50 miles per hour. The winds on one side will also be traveling north. To figure out how fast the wind will be on the ground in that part of the tornado, we have to add the speed of the tornado as a whole—50 miles per hour—to the circular speed of its winds—150 miles per hour. At that point on the ground, you would feel winds of 200 miles per hour.

The winds on the opposite side, however, are traveling in the opposite direction, and so the two forces partly cancel each other out. The wind speed on the ground in that side of the tornado will be much lower, only 100 miles per hour.

Most tornadoes turn counterclockwise, and so as the tornado advances, the highest winds will be on the left-hand side as you face it. Or, to look at it another way, since most tornadoes in North America travel northeast, the highest winds are most often on the southeast side of the tornado.

Ahrens, C. Donald. *Meteorology Today: An Introduction to Weather, Climate, and the Environment.* St. Paul: West Publishing Company, 1991.

Sickle-cell Anemia

Anemia develops when the body either can't produce enough red blood cells, or produces defective cells instead. Because red blood cells are responsible for carrying oxygen from the lungs to the rest of the body, people with anemia feel weak and have difficulty exerting themselves.

Sickle-cell anemia is a dangerous disease with an interesting side benefit. Dietary deficiencies are responsible for most forms of anemia, but sickle-cell anemia is hereditary and occurs most commonly among people of West African descent. Sickle cells are red blood cells with a slightly different chemical composition from normal red blood cells. When the oxygen level in the blood drops too low, the defective cells crystallize into sickle-cell shapes. In this condition the blood cells can no longer transport oxygen. The result for the person affected is shortness of breath, stomach pains, internal bleeding, and sometimes even blood clotting.

An individual may inherit either one or two genes for sickle-cell. True sickle-cell anemia requires two genes—one from each parent. With only one gene, the condition is called "sickle-cell trait." Sickle-cell trait is still dangerous at high elevations, or in other situations involving low levels of oxygen, but it is rarely fatal and is usually not even noticeable.

In Africa, sickle-cell trait carries a distinct advantage. The chemical composition of the sickle cell prevents it from being invaded by malarial parasites, making people with sickle cells immune to malaria. In West Africa, where malaria is a problem, as much as 20 percent of the population has sickle-cell. In the United States, where malaria is not a problem, sickle-cell offers no advantage. Here, fewer than 10 percent of African Americans carry genes for sickle-cell. And the incidence of sickle-cell anemia is even lower, at only 1 in 400.

Rust

Water and oxygen are both necessary for metal to rust, but equally important is the flow of electrons from the metal into the surrounding water. In order for iron to be dissolved in water, it has to have some kind of electrical charge. All atoms begin with a balanced number of negatively charged electrons and positively charged protons. So when the iron gives up electrons, it then has a net positive charge.

Once the iron is positively charged, it can react with oxygen dissolved in the water to form iron oxide, or rust. As the rust forms, the iron loses its positive charge. Now it can give up more electrons to the water, and so the cycle continues.

Usually rust forms at two different sites in two different ways. For example, a spike stuck partway into the ground rusts fastest underground, where the metal can give up electrons into the moist soil. As the spike gives up electrons, the whole spike becomes positively charged. When that happens, the metal at the top of the spike, which is exposed to the oxygen in the air, also forms a coating of rust.

We've seen how metal exposed to salt, whether it be seawater or salt put on the roads in winter to melt snow and ice, rusts faster. Salt speeds up corrosion because salt water conducts electricity better than fresh water, making it easier for iron to give up electrons.

In the 1930s, scientist Michael Faraday was the first to explain metal corrosion as an electrochemical process. Since then we have learned a lot more, but there is still a great deal that is not known about how rust works. A more recent wrinkle is that scientists have found that bacteria, too, are involved in metal corrosion, making it a biological process as well. So if you drive an old car, it may have even more bugs in it than you think. Rust involving bacteria is called "biocorrosion."

Bacteria can work to either slow down or speed up rust by creating what is called a "biofilm" over the metal. This biofilm, made up of living bacteria, can act like a coat of paint or rustproofing, protecting the metal from the corrosive effects of moisture.

But the biofilm can also have the opposite effect if other types of bacteria get to the metal first. For example, sulfate-reducing bacteria draw dissolved sulfur out of the water and release sulfuric acid. The sulfuric acid is extremely attractive to electrons, and causes the metal to corrode much faster than normal. But sulfate-reducing bacteria work better when there is no oxygen present. If the sulfate-reducing bacteria get to the metal before the biofilm forms, the film can actually cover and protect the damaging bacteria from exposure to oxygen and water currents. Sealed under the biofilm, the sulfate-reducing bacteria can work away at the metal relatively undisturbed.

"Metals, Corrosion of." In *Collier's Encyclopedia*. New York: Macmillan Educational Company, 1987.

Sienko, Michell J., and Robert A. Plane. *Chemistry*. 5th ed. New York: McGraw-Hill, 1976.

How Can You Tell If a Spider Is Dead?

If a spider is not moving and all its legs are flexed—that is, pulled in toward its body—it's likely to be dead. Although spiders' legs have flexor muscles—muscles that bend the legs in toward the body—they do not have extensor muscles—muscles that would cause the legs to straighten and point away from the spider's body.

So a spider flexes its legs by using its flexor muscles. But how does a spider extend its legs? In the 1940s the zoologist C. H. Ellis noted that, as a rule, dead spiders have flexed legs. Evidently, whatever straightens a spider's legs in life is inoperative in death.

Ellis and other zoologists demonstrated that spiders extend their legs with a hydraulic system. The legs of a living spider contain fluid under pressure that tends to straighten the legs, just like water pressure stiffening a garden hose, or hydraulic fluid pressure lifting a car at a garage. The spider increases fluid pressure when it wants to extend its legs more forcefully. If a spider's leg is cut, the spider can't straighten that leg until it seals off the fluid leak.

If the spider dies, it can't maintain its internal fluid pressure. The leg flexor muscles may contract one more time, but without fluid pressure there will be no opposing force to straighten the legs again. That's why a motionless spider with flexed legs is likely to be dead.

Vogel, Steven. *Life's Devices: The Physical World of Animals and Plants.* Princeton, N.J.: Princeton University Press, 1988.

Degradable Plastics

Past ages have been called the Stone Age, Iron Age, and Bronze Age. Future archaeologists may call this the Plastic Age.

Plastics are widely used in part because they can't be broken down by natural agents, as can some of the material they have replaced. But the fact that plastics don't break down is troublesome. Plastic litter outlasts litter made of natural materials. So as we run out of space in landfills, plastics become more and more of a problem.

There is no perfect solution, but we may find some help from a new and seemingly contradictory technology: degradable plastics.

Biodegradable trash bags combine starch and plastic. When bacteria eat the starch, much of the bag then turns to dust. The bag still doesn't break down completely, but microorganisms can at least get to the garbage inside, which otherwise is isolated for as long as the trash bag holds together.

Photodegradable plastic breaks down after a few months' exposure to the sun. Photodegradable plastic won't help with trash buried under a landfill, but it could help with litter.

In the long run, recycling may solve more of our problems with plastic waste. But plastic is harder to recycle than glass or metal, partly because there are many different kinds of plastic with different chemical properties.

Ultimately, degradable plastic presents a dilemma. If we make plastics that can be broken down by the sun or microorganisms, some of the benefit of plastic is lost. Since many of the products made from recycled plastic have to withstand the forces of nature, degradable plastic could make recycling *more* difficult. Fences and park benches, for example, can be made from recycled plastic, but if either contains photodegradable material, it won't last long in a sunny park.

13

Beardsley, Tim. "Disappearing Act: Can Degradable Plastics Ease the Landfill Crisis?" *Scientific American,* November 1988.

A Molecular Soccer Match

In 1827, botanist Robert Brown noticed tiny particles of plant pollen jiggling randomly around in a dish of water under his microscope. Brown knew that the particles weren't alive, but he couldn't explain their movement.

Nearly a century later, Einstein used Brown's observations as evidence in one of the major debates of his time—the question of whether or not molecules existed. One group, including Einstein, believed that matter was made up of smaller particles like sand on a beach. Other scientists claimed that matter was continuous like a smooth slab of rock. No one could see molecules, but Einstein argued that Brown's particles moved as they did because they were being hit by water molecules.

To see how, imagine a field of people, all pushing an enormous ball. Each time someone hits the ball, it moves a tiny bit. The ball moves this way and that, and gradually works its way around the field. A distant observer can't see individual people, and so the ball appears to move randomly on its own.

The tiny particles drifting in water are much bigger than the invisible water molecules, but the molecules push the particles around the way the crowd pushes the ball. Physicists have named this random jiggling motion "Brownian movement," after the botanist who was astute enough to observe the strange motion, even though he had no idea of its cause.

Little did Brown know that in 1905 a young physicist just starting his career would use those jiggling pollen particles to answer one of the biggest questions of physics at the turn of the century.

Feynman, Richard P. *The Feynman Lectures on Physics.* Reading, Mass.: Addison-Wesley, 1964.

14 *Conversation at a Crowded Party*

"I don't know what she sees in him."
"Beg your pardon?"
"I say, I DON'T KNOW WHAT SHE SEES IN HIM."

An article published in 1959 in the *Journal of the Acoustical Society of America* presented a rough theoretical analysis of sound levels at cocktail parties. The author, William MacLean, analyzed the problem of carrying on a conversation in the presence of background noise from other conversations, and he made a prediction that you can check for yourself.

At the beginning of a party, when few guests are present, quiet conversation is possible. As more guests arrive, you have to talk louder and louder to override the increasing background noise.

MacLean's calculations predicted that when the number of guests at a party exceeds a certain maximum determined by the size and other characteristics of the room, merely talking louder is of no avail in continuing your conversation. You just force everybody else to talk louder. The ensuing increase in background noise soon drowns you out unless you move closer to the person you're talking to—closer than you might get in another situation.

The acoustics of real rooms are so complex that it's practically impossible to say exactly when this need to get closer will set in—but MacLean predicted that that moment will occur at some point as more and more people arrive.

Someone may temporarily quiet a loud party, perhaps to introduce the guest of honor. But, MacLean found, even if everyone tries to talk quietly afterward, dialogues like the one we began with eventually drive the background noise back up to its earlier level. A crowded party remains loud until guests begin to leave.

Hall, Edward T., Jr. "The Anthropology of Manners." *Scientific American,* April 1955.
MacLean, William R. "On the Acoustics of Cocktail Parties." *Journal of the Acoustical Society of America,* January 1959.

Bloodletting 15

If you got sick 200 years ago, your doctor might well have drawn out some of your blood. Bloodletting hasn't been common in Western medicine for more than a hundred years, but some studies may have found medical evidence for the value of this practice.

All living cells, whether they are part of a person or of a bacterium, need iron to live. In order to meet this need for iron, everything from drink mixes to breakfast cereals is now fortified with iron. Lots of breakfast cereals advertise that one helping contains all the iron you need for that day. That means that if you eat any more iron that day, you'll be getting more than you need.

But the bacteria that make us sick also need iron, and one of the ways the body fights disease is to lower its own level of iron in order to starve the bacteria. After surgery, during the growth of cancer cells, or whenever there is a threat of disease or infection, the concentration of iron in the body goes down. And studies have shown that giving extra iron to people in these situations may actually increase the danger of disease or infection by making the bacteria healthier.

When the old-time doctors drew blood, the idea was to get rid of poisons in the blood. In fact, their method may have worked because they were getting rid of valuable nutrients that would otherwise have strengthened the microorganisms causing the disease. Most of the body's iron is in the blood, and so by getting rid of the iron, the doctors may have been slowing the growth of the disease.

Extremely low iron levels are just as dangerous, and no one yet is advocating a large-scale return to bloodletting. But it looks like we may be taking another look at one old medical practice, once chalked up to ignorance.

Kent, Susan; Eugene D. Weinberg; and Patricia Stuart-Macadam. "Dietary and Prophylactic Iron Supplements: Helpful or Harmful." *Human Nature* 1, no. 1 (1978).
Weinberg, Eugene D. "Iron Withholding: A Defense against Infection and Neoplasia." *Physiological Review* 64 (1984).
Weinberg, R. J.; S. R. Ell; and E. D. Weinberg. "Blood Letting, Iron Homeostasis, and Human Health." *Medical Hypotheses* 21 (1986).

16

Looming on the Horizon

When we say that distant mountains are looming on the horizon or, more figuratively, that an important event is looming, we are using an old sailor's term for a particular kind of mirage.

One way to see looming is to stand on the shore of a large lake or sound on a sunny afternoon and look at the horizon, preferably through binoculars. If you see the water surface at the horizon appearing to curve upward, giving you the impression that you are inside a shallow bowl, you are seeing the kind of mirage known as looming.

On a sunny afternoon, warm air is likely to be moving from the land out over the cool water, which cools that air from below. Looming arises from the bending of light rays as they leave the distant water surface and pass through that cool layer of air near the surface into warm air higher up.

Whenever light rays pass from a dense medium like cool air into a less dense medium like warm air, the rays are bent. In the case of looming, light rays leaving the distant water at a shallow angle that would otherwise cause them to pass over your head unseen are instead bent downward, toward the horizontal, as they travel from cool air into warm air.

When those light rays, now traveling horizontally, reach your

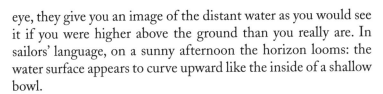

eye, they give you an image of the distant water as you would see it if you were higher above the ground than you really are. In sailors' language, on a sunny afternoon the horizon looms: the water surface appears to curve upward like the inside of a shallow bowl.

Fraser, Alistair B., and William H. Mach. "Mirages." *Scientific American,* January 1976. Reprinted, with introductions by David K. Lynch, in the *Scientific American* anthology *Atmospheric Phenomena* (1980).

A Wet Paintbrush

Take an artist's watercolor brush, dip it in water, and pull it out again. The bristles cling together to form a smooth, pointed shape that artists and calligraphers use to paint lines of varying thickness. Something similar happens when a person with straight hair dives into water and climbs out again: that person's hair is slicked down.

We usually say that the bristles or hairs cling because they are wet. But that can't be the real explanation, as you can see if you look at the bristles of that brush while you hold it underwater. While immersed in water, the bristles do not cling, even though they are certainly wet. Wet bristles—and wet human hairs—cling together not if they are surrounded by water, but only if they are surrounded by a water *surface.*

The clinging of the bristles is really a manifestation of the clinging of individual water molecules. A water molecule at the surface of a body of water—say, on the outside of a wet brush—is pulled strongly toward that body of water, because that's where the other water molecules are. One result of this mutual attraction between water molecules is that a water surface is under tension, like an elastic skin.

That surface tension pulls water into beads on a well-waxed car. It also holds the bristles of a wet brush together—if the brush is surrounded by air. The bristles of a brush immersed in water don't cling because they are not surrounded by a water surface.

Boys, C. V. *Soap Bubbles and the Forces Which Mould Them.* Garden City, N.Y.: Doubleday Anchor Books, 1959.

Putting on the Brakes

A dog runs in front of your car, and your foot jumps to the brake. But for the quickest stop, should you lock the wheels and skid to a stop, or brake more gently so the wheels still turn? Either way, what stops the car is the friction of the tires on the road—the greater the friction, the quicker the stop. Part of the complexity of this problem is that there are two different kinds of friction: sliding and static.

First, let's take a different situation involving friction. Imagine that you are trying to push a heavy box down a wooden ramp. At first it won't move, because on a microscopic level the two surfaces are rough enough to catch on each other and resist your push. The rough surfaces conform like two open egg cartons stacked together. The force that the two rough surfaces exert to hold the box in place is called *static friction*.

But when the box does start to move, you need less work to keep it going. That's because when the surfaces slide they separate slightly, and the higher spots on one rough surface move across the higher spots on the other surface. The rough surfaces still exert some force, slowing the box down; this force is called *sliding friction*. And, as you can tell from your own efforts with the box, static friction can exert more force than sliding friction.

But how does this apply to the car? As long as the wheels turn, some part of the tire is always planted firmly on the road. As you brake, the momentum of the car forces the tires forward against the road surface, just as when you pushed on the box. The harder you brake, the more static friction develops between the road and the tires—until you start to slide. At that point, the sliding friction offers less resistance to the car's momentum than the static friction did just before you started to skid. You can stop the fastest, then, by not quite skidding, and that is the principle behind antilock brakes.

Walker, Jearl. "The Amateur Scientist." *Scientific American,* February 1989.

Good Science, Bad Results

In the nineteenth century, geology was a new science. And as geologists studied the earth's rocks and fossils, they began to suspect that the planet was far older than anyone had imagined. Until about the middle of the eighteenth century, scholars had set the beginning of the earth at between 4,000 and 5,000 B.C. But by the middle of the nineteenth century, a hundred years of geological discoveries had extended the age of the earth from a mere 6,000 years into the billions of years. Then, in 1868, the British physicist Lord Kelvin used the earth's temperature to calculate a much shorter age for our planet.

Most geologists accepted that the earth had started as a molten mass and had been cooling ever since. Kelvin calculated how long it would take the earth to cool to its current temperature from its original molten temperature. Volcanoes proved that the inside of the earth was extremely hot. And, he reasoned, for the earth still to be that hot, it could not be more than 100 million years old, and perhaps as young as 26 million. The age of the earth according to Kelvin's calculations was short enough to force major changes in the way nineteenth-century scientists imagined the world.

No one at the time could refute Kelvin's results because he accounted for all the known laws of physics, and he used evidence that any geologist could measure. Kelvin's calculations were so simple and straightforward that today some historians wonder why no one had presented them earlier.

Kelvin was wrong, but not because of his calculations. Early this century, physicists discovered that naturally occurring radioactivity in the earth was releasing enough heat to offset much of the earth's cooling. Taking into account this radioactivity, geologists now estimate the age of the earth at 5 billion years.

Bowler, Peter. *Evolution: The History of an Idea.* Berkeley: University of California Press, 1984.

Gould, Stephen Jay. *Time's Arrow, Time's Cycle.* Cambridge, Mass.: Harvard University Press, 1987.

Hawking, Stephen. *A Brief History of Time.* New York: Bantam Books, 1988.

Tickling the Funny Bone

Of all the parts of our body, the funny bone may have the least appropriate name. That sensitive point on the elbow is not a bone, and it certainly isn't funny.

Most of us have at some time banged an elbow on a sharp corner and felt that indescribable tingling—like an electric shock—up and down the arm. What we are hitting is not a bone but a bundle of nerves, causing them all to fire at once.

Thousands of nerves carry messages from every part of the arm to the brain. Some report on heat, others on cold, and others on pressure. One bundle of nerves passes through a channel in the elbow that we call the funny bone. If you could intercept some of the messages along the way, the messages would all look the same—a combined electrical and chemical impulse. Your brain recognizes heat, cold, and pain only because it knows which nerves sent the signal.

The nervous system is generally reliable, but sometimes it can fool us. Amputees, for example, who have lost a leg may feel pain in the missing foot. The foot is gone, but if the nerves that connected the foot to the brain fire, the brain interprets the signal as pain in the foot.

When you bang the nerves passing through your funny bone against the corner of a table, the shock causes all the nerves to send their messages at the same time. So the message the brain gets is a confused combination of cold, pain, heat, and everything else which it interprets as coming from all over the arm.

A jolt of electricity can also cause nerves to fire randomly, and that is why a bump on the funny bone feels like an electric shock. So banging the funny bone is not very funny, but the reason it feels the way it does is that your brain doesn't know *what* to think.

Putting South on Top

Maps of the world usually have north at the top and Greenwich, England, on a north-south line which runs through the center of the map. Not quite all world maps are laid out this way, however. In the mid-1980s an Australian cartographer by the name of McArthur published his so-called "McArthur's Universal Corrective Map of the World."

McArthur's map looks almost like one of the rectangular world maps we see all the time, but turned upside down. Also, Australia, not England, is at the horizontal center. The printed words on the map—the title and the names of oceans, countries, and a hundred or so major cities—are right side up only if you orient the map with south at the top. Looking at McArthur's world map is a strange experience. The United States occupies the lower left corner, with the Pacific Ocean to the right of North America and the Arctic Ocean below.

McArthur says that his map begins a crusade against "the perpetual onslaught of 'downunder' jokes—implications from Northern nations that the height of a country's prestige is determined by its equivalent spatial location on a conventional map of the world."

McArthur is only half joking. Cartographers remind us that any map is merely an interpretation of the world, that no single map can serve all purposes, and that maps shape and are shaped by the self-images of the people who make and use them.

"New Map Turns World Geography Upside Down." *Earth Science,* Fall 1985. (This article says that 35" x 23" laminated copies of McArthur's map can be bought [or could be bought in 1985] from Rex Publications, 413 Pacific Highway, Artarmon 2064, Australia.)

Phipps, William E. "Cartographic Ethnocentricity." *The Social Studies,* November–December 1987.

Porter, Phil, and Phil Voxland. "Distortion in Maps: The Peters Projection and Other Devilments." *Focus,* Summer 1986.

Mayonnaise Emulsions

When we say that two people are like oil and water, we usually mean that they don't get along together. Oil won't mix with water, or vinegar, and that's why you often have to shake salad dressing before you use it.

But there is a way of mixing oil and vinegar so they stay mixed. One such mixture is mayonnaise. To make mayonnaise, you need to make what's called an emulsion. An emulsion combines two liquids that ordinarily don't mix by breaking one of the liquids up into tiny droplets that stay suspended in the other liquid.

Try mixing oil and vinegar by shaking them in a jar. When you stop, you can see that the oil is broken up into tiny particles, swirling about in the vinegar. But if you want to use the mixture on a salad, you'd better hurry, because almost immediately the oil droplets float to the surface and separate from the vinegar.

An emulsion, such as mayonnaise, needs an emulsifying agent to keep the droplets in suspension. In mayonnaise, egg acts as the emulsifying agent. To make mayonnaise you beat egg yolks with a little lemon juice, then add oil very slowly as you continue beating the mixture. The beater not only breaks the oil into tiny droplets, just as you did when you shook the jar, but also coats the droplets with egg yolk. The yolk coating prevents the droplets from coming back together, and so is said to "emulsify" the oil.

When you finish making mayonnaise, you'll probably want to use soap to clean out the oily dishes. Soap is also an emulsifying agent. As you scrub the dishes, the soap wraps itself around the oil, drawing it into suspension in the dishwater so it can be washed away.

The Joy of Cooking has some easy recipes for mayonnaise. Try experimenting with different varieties and amounts of oils. You'll not only learn about emulsions, but make better sandwiches, too.

McGee, Harold. *On Food and Cooking: The Science and Lore of the Kitchen.* New York: Scribner, 1984.

Learning to Talk

Parents of young children are sometimes surprised when their preschoolers' grammar starts getting worse. When children begin using sentences, their grammar is reasonably correct. Then, as they get a little older, the same children start making more mistakes. A child who used the word "mice" might start saying "mouses," even though she had never heard anyone else make that mistake. If you see this pattern in your own child, don't worry; child development specialists say it's actually a good sign. By making these mistakes, children are showing a deeper under-standing of grammar.

The early sentences are grammatically correct only because the child is imitating adults. When you say, "We went to the zoo," the child repeats, "We went to the zoo." But a year later, when you ask, "Did you go outside?" the same child might answer, "Yes, I goed outside." To say "I goed outside" may not be correct, but it does follow the common rule that the English past tense is usually formed by adding *-ed*. In this later stage of language learning, the child is breaking up phrases such as "We went outside" into words and trying to recombine them to mean different things. The collection of rules that we use to combine words in different ways is grammar.

We think we teach children grammar by correcting their mistakes, but a child learns about grammar mostly through listening to others, and by trial and error. No one told the child to say "mouses" for "mice," or "goed" for "went"; those words children figure out on their own based on the English grammar rules that form plurals by adding *-s*, and past tenses by adding *-ed*. And just as children learn the mistakes by themselves, they eventually learn the correct forms by themselves as well.

Baby's First Steps

Children don't learn to walk until about age one. But if you hold a newborn baby so its feet just touch the ground, an automatic response causes the baby to lift its legs as if walking. What is more puzzling, though, is that this automatic stepping response stops after a few months, and doesn't return until about six months later.

Scientists have offered several complicated explanations. One suggestion was that the newborn's stepping response was a primitive instinct which disappeared as the baby developed, and that the later stepping was a more conscious, human effort at real walking.

But the complicated, theoretical explanation is not always the correct one. Indiana University psychologist Esther Thelen used two simple experiments to demonstrate a much simpler answer. Thelen's explanation was this: The fat in a baby's legs increases faster than the baby's muscular strength. When the legs get too heavy, the baby stops trying to lift them. In other words, a six-month-old baby doesn't try to walk when its feet touch the ground because its legs are too fat.

Thelen's experiment involved a group of babies who had grown out of the automatic stepping stage. She held these babies in shallow baths of water so that the buoyancy of the water would take most of the weight off their legs. When their feet touched the bottom, the babies began stepping.

In the other experiment, Thelen tied tiny ankle weights to the legs of a group of newborn babies, and held them in the air so their feet just touched the ground. The weights acted like the added fat in older babies, and the newborns did not begin stepping.

The two experiments showed that whether or not a baby picks up its feet when they touch the ground depends on the weight of the baby's legs. Thelen's practical experiments solved a problem that had puzzled developmental biologists for a long time.

Whole-Wheat Bread Is a Mixed Bag

One of the ideas behind the interest in "whole" foods, including whole-wheat bread, is that if you eat more of the plant, you get more of the vitamins. When the outer layer is removed from whole wheat to make white flour, the flour does lose iron and some B vitamins, but by leaving that layer on you may be losing other nutrients.

Although it might seem improbable that by taking part of the grain away you would get more of some nutrients, calcium is one important mineral that is more available from white flour than from whole-wheat. Whole-wheat flour has just as much calcium as white flour, but whole-wheat also contains phytic acid, which binds up the calcium, making it harder for your body to absorb.

If you get enough calcium in other forms, such as milk, you don't need as much calcium from bread, but for people who already have inadequate diets—and especially for children with growing bones—the loss of calcium can be serious. When Dublin was put on a diet of whole-wheat bread during World War II, the result was a massive epidemic of the bone disease rickets.

In a healthy diet, the main advantage of whole-wheat bread is the fiber, which is undigestible but is still necessary for good digestion. By adding bulk, and speeding up the digestive process, the fiber from whole-wheat can help prevent appendicitis, gallstones, hardening of the arteries, and some types of cancer.

But, like all food, whole-wheat bread is a mixed bag. For most of us, the benefits of the fiber probably outweigh the slight loss of nutrients, but the value of any food depends as much on the other things you eat as it does on the chemistry of that particular food.

McGee, Harold. *On Food and Cooking: The Science and Lore of the Kitchen.* New York: Scribner, 1984.

Packaging for the Birds

If you were manufacturing a highly perishable product, you might not want a porous container to pack it in. And yet that is exactly what hens have been doing for thousands of years. In order for a chick embryo to develop, the eggshell has to be slightly porous so that oxygen can pass through. But the danger of a porous shell is that bacteria can get through as well.

The catch is that for bacteria to get to the yolk and the unhatched chick, they have to pass through the white of the egg, where the chicken has set a chemical trap for unwanted cells. Bacteria, like all animals, need iron to survive, and the unhatched chick gets its iron from the yolk. But the white that separates the iron-rich yolk from the shell has virtually no iron. And not only is the white low in iron, it also contains a protein called conalbumin that is strongly attracted to iron. Because there is far more conalbumin than iron in egg white, the conalbumin kills any cells that come through the eggshell by stealing their iron. Eventually the conalbumin gets enough iron that it's no longer attracted to the iron in the bacteria, and the egg will spoil.

When the Earl of Gloucester in Shakespeare's *King Lear* is

injured, a servant shouts, "I'll fetch some flax and whites of eggs to apply to his bleeding face" (Act III, Scene 7). No one had heard of conalbumin in the seventeenth century, but the servant knew that egg white would prevent the infection of his master's eyes. Chickens probably don't know about conalbumin either, but their eggs demonstrate the efficiency of one natural immune system that scientists are just beginning to understand.

Kent, Susan; Eugene D. Weinberg; and Patricia Stuart-Macadam. "Dietary and Prophylactic Iron Supplements: Helpful or Harmful." *Human Nature* 1, no. 1 (1978).
Weinberg, Eugene D. "Iron Withholding: A Defense against Infection and Neoplasia." *Physiological Review* 64 (1984).

Look through Your Comb at the Mirror

26   Hold a pocket comb, with teeth vertical, between your eyes and a bathroom mirror. Look through the teeth of the real comb at the teeth of the reflected comb. Slowly move the comb toward the mirror, always keeping both the comb and its reflection in your line of sight. As the comb gets within a few inches of the mirror, you will see what appears to be a shimmering, magnified view of the comb's teeth, with the magnifying power steadily increasing as the comb approaches the mirror.

The shimmering image is a pattern of light and dark created by overlap of the teeth of the real comb and the reflected comb. In some places, you see the teeth of the reflected comb fill in gaps between the teeth of the real comb, and you see solid black. In other places, gaps between teeth on the real comb line up with gaps in the reflected comb. Those areas appear relatively light.

The eerie pattern of dark and light areas is an example of a so-called moiré pattern. Moiré patterns arise whenever two repetitive gridlike designs overlap. In this case, the repetitive design is the row of evenly spaced teeth on the comb. And moiré patterns often resemble a magnified view of the overlapping designs. For example, when you look at overlapping folds of sheer drapery fabric, you see a moiré pattern of crisscrossed dark lines that look like a magnified view of the fabric.

Returning to our comb example: Notice that the magnified teeth in the moiré image are even tapered, just like the real ones, and that if you point the teeth slightly toward or away from you, the moiré image appears to do the same thing. You have to see it to believe it.

Stecher, Milton. "The Moiré Phenomenon." *American Journal of Physics*, April 1964.

Opera Singers Cut through the Orchestra

First-class opera singers can make themselves heard distinctly even over a fairly loud orchestra. Acousticians have found that singers accomplish this at least partly by making extra sound at certain moderately high frequencies where the orchestra is not especially loud.

Recall that the sound of a singing voice or a musical instrument is really a complex mixture of vibrations at many different frequencies. Each instrument and each voice has its own peculiar mixture of frequencies that we perceive as its tone color.

Good opera singers learn, by one method or another, to produce a tone that contains an especially large amount of sound energy in the frequency range of 2,000 to 4,000 vibrations per second. The emphasis on frequencies in the 2,000-to-4,000-vibration-per-second range makes the operatic voice sound very different from the pop singer's voice and from ordinary speech.

The sound of a symphony orchestra, on the other hand, does not have any special emphasis on frequencies between 2,000 and 4,000 vibrations per second. So opera singers cut through the orchestra by emphasizing frequencies that the orchestra does not.

Pop singers, by the way, often don't emphasize those special high frequencies because they want to create a more conversational, speechlike sound than an opera singer. But at least one respected book on sound recording recommends that the pop recording engineer electronically amplify the frequencies between 2,000 and 4,000 vibrations per second on vocal tracks to keep the voices from being buried by the accompanying instruments!

Runstein, Robert E., and David Miles Huber. *Modern Recording Techniques.* Indianapolis: Howard W. Sams and Company, 1986.
Sundberg, Johan. "The Acoustics of the Singing Voice." *Scientific American,* March 1977. Reprinted in the *Scientific American* anthology *The Physics of Music.*

The Roots of "Algebra"

Our word "algebra" comes from one word in the long Arabic title of a mathematics book written about 1,150 years ago in Baghdad. The author was the Islamic mathematician and astronomer Abu Jafar Muhammad Ibn Musa al-Khwarizmi. The Arabic word in question, "al-jabr," was used by al-Khwarizmi to mean restoration or completion—one of the techniques he recommended for solving an equation.

In modern language, this is the technique of taking terms preceded by a minus sign on one side of an equation and moving them to the other side, where they can then have plus signs. Maybe you remember learning this trick in school.

The full title of al-Khwarizmi's *Algebra* has been translated as *The Compendious Book on Calculation by Completion and Balancing.* The book's stated purpose was to provide "what is easiest and most useful in arithmetic, such as men constantly require in cases of inheritance, legacies, . . . lawsuits, and trade." Al-Khwarizmi's *Algebra* was a standard text in the Middle East and Europe well into the Middle Ages.

Unlike modern algebra books, al-Khwarizmi's *Algebra* used neither x's nor y's, nor even numerals. All the problems were stated in words; for example: "A quantity: I multiplied a third of it and a dirham by a fourth of it and a dirham; it becomes twenty."

Al-Khwarizmi did use numerals in some of his other works, including a book explaining arithmetic with Hindu numerals. Those Hindu numerals were the ancestors of the numerals 1 through 9 and zero that we use today. But al-Khwarizmi's reputation was so great that those symbols later became known as Arabic numerals. And al-Khwarizmi's name is the root of the modern word "algorithm," meaning a step-by-step procedure for solving a mathematical problem.

"Al-Khwarizmi." In *Dictionary of Scientific Biography,* Charles Coulston Gillispie, Editor-in-Chief. New York: Scribner, 1971.

Morality and Nutrition

Health food is popular in this country—mostly for the sake of our physical health. A hundred years ago, around the end of the last century, the United States was in the middle of its first health-food craze, but back then physical health was only part of the reason. Refined foods, additives, and stimulants were considered immoral as well as unhealthy.

In a book called *Plain Facts for Old and Young,* Dr. John Harvey Kellogg wrote, "A man who lives on pork, fine-flour bread, rich pies and cakes, and condiments, drinks tea and coffee, and uses tobacco, might as well try to fly as to be chaste in thought." The message was that if you eat impure food, you'll have impure thoughts.

The center of the health-food movement in the nineteenth century was Battle Creek, Michigan, which was also the world headquarters of the Seventh-day Adventist church. In 1863, Ellen White, who was then the leader of the church, had a religious revelation telling her to eat only certain foods. As a result, meat and stimulants became forbidden by church doctrine, and White hired Dr. Kellogg to run the Seventh-day Adventists' sanatorium in Battle Creek. In order to make the diet more interesting, Dr. Kellogg developed a new breakfast cereal by baking oats, wheat, and cornmeal into hard biscuits and then crumbling the biscuits into small chunks. He named his new cereal "granola" because it was made from granules of whole-grain biscuits. So the next time you sit down to a bowl of granola, remember that it was invented not just for your physical health, but for your moral health as well.

McGee, Harold. *On Food and Cooking: The Science and Lore of the Kitchen.* New York: Scribner, 1984.

Moral Fiber in Whole-Wheat Bread

What do you say when the person behind the sandwich counter asks you, "Would you like that on white or whole-wheat?"

Actually, the interest in whole-wheat bread is relatively new.

For thousands of years, wheat has been ground and the non-digestible outer layer thrown away. Until the nineteenth century, when machinery made it easier to process flour, white flour was much more expensive than whole-wheat. The cost, along with the lighter color and texture, made white bread a status symbol.

But the first large movement to encourage the use of whole-wheat bread was not based solely on physical health. In 1837 a Presbyterian minister named Sylvester Graham wrote *A Treatise on Bread and Bread-Making,* in which he denounced white bread as unnatural, since he said God had made wheat with both a digestible and a non-digestible part.

Today, health experts recommend the non-digestible fiber of whole-wheat bread because it adds bulk and speeds up the digestive process. Nobody knows for certain how much dietary fiber really helps, but it is thought that as the food passes more quickly through the intestines, there is less time for the body to absorb toxic chemicals.

Although Graham didn't have the information that we have today, he recognized the importance of the non-digestible part of the wheat. Only according to Graham, the benefits of whole-wheat bread included not only physical health but also, in his words, the "intellectual, and moral, and religious, and social, and civil, and political interests of man."

Now, our reasons for eating whole-wheat bread may be a bit more modest, but Sylvester Graham was the first person known to have argued publicly for the benefits of dietary fiber. Even today, whole-wheat flour is sometimes called "graham flour," after Sylvester Graham. And graham crackers are another familiar food named after the Reverend Sylvester Graham.

McGee, Harold. *On Food and Cooking: The Science and Lore of the Kitchen.* New York: Scribner, 1984.

Blow Out Candles with an Oatmeal Box

To do this trick, you need an empty cylindrical cardboard oatmeal box with its lid. Cut a round hole the size of a penny in the center of one end of the box.

Now aim the box at a lighted candle, with the hole facing the candle. Tap sharply on the other end. If you have aimed the oatmeal box correctly, the candle will be suddenly blown out a moment after you strike the box. With some practice you can blow candles out from up to six feet away.

When you strike the box, a so-called vortex ring comes out of the hole. This vortex ring is a region shaped like a rubber O-ring constantly turning itself inside out. That turning-inside-out motion of the air enables the vortex ring to retain its shape as it travels toward the candle. That same motion, combined with the forward motion of the vortex ring, blows the flame out when the ring arrives at the candle.

You can make that vortex ring visible as a smoke ring. Fill the oatmeal box with cigarette smoke, then tap very gently on the end of the box. A smoke ring will emerge from the hole, travel relatively fast for a foot or two, then slow down and spread out. If you tap harder, the smoke ring will travel farther but will be harder to see.

Beeler, Nelson F., and Franklyn M. Branley. *Experiments in Science.* Revised enlarged edition. New York: Crowell, 1955.

Feynman, Richard P. *The Feynman Lectures on Physics.* Reading, Mass.: Addison-Wesley, 1963.

What Temperature Boils Down To

Water boils at 212 degrees and freezes at 32. The human body temperature is about 98.6 degrees, and a comfortable room is between 65 and 70. We take these kinds of numbers for granted, but have you ever wondered how such seemingly arbitrary numbers came to have such meaning?

Actually the numbers are arbitrary. If you wiped all the numbers off of a thermometer and replaced them with your own, you could still use that thermometer to compare the temperature in your house from one day to the next. But you wouldn't be able to compare the temperature of your house with the temperature of another house down the street.

The first widely accepted temperature scale—and the one most commonly used in the United States—is the Fahrenheit scale developed by the Dutch physicist and instrument-maker Gabriel Daniel Fahrenheit in the early 1700s. When Fahrenheit was working, there were at least 35 different temperature scales. So if someone at that time said the temperature was 65 degrees, you still wouldn't know whether it was hot or cold unless you knew what scale the temperature referred to.

Fahrenheit assigned the number zero to the lowest temperature that he could get by mixing salt and ice. He then assigned the number 96 to body temperature. Using his new scale, Fahrenheit found that pure water freezes at 32 degrees and boils at 212 degrees. As temperature measurements became more accurate, Fahrenheit's scale remained in place, even though physiologists figured out that typical human body temperature is about 98.6 degrees Fahrenheit and not 96, as Fahrenheit had calculated.

The United States is one of only a few countries that use the Fahrenheit scale. Most others use a scale developed by the Swedish astronomer Anders Celsius about a decade after Fahrenheit developed his. The Celsius scale was created by assigning the number zero to the temperature at which water freezes, and 100 to the temperature at which it boils. Speaking in Celsius, a healthy body temperature is around 37 degrees, and an average room is about 20 degrees.

On the Fahrenheit scale, there are 180 degrees between the freezing point—32—and the boiling point—212. On the Celsius scale, there are only 100 degrees between the freezing and boiling points of water. That means that each Celsius degree is 1.8 times the size of a Fahrenheit degree. So a temperature change of 10 degrees Celsius is the same as a change of 18 degrees Fahrenheit.

Normally Americans don't have too many occasions to convert between Fahrenheit and Celsius, but if you can think of 20 degrees as a comfortable room temperature, and 40 degrees as a very hot day, then when someone tells you it's a balmy 24 degrees out, you'll know what they mean without converting at all.

Both the Fahrenheit and Celsius scales are *relative* scales because they tell us the temperature relative to some other temperature. But for most other measurements we use *absolute* scales. If you start with a 25-pound sack of flour and take out 10 pounds, you'll be left with 15. If you take away another 15 pounds, you will have zero pounds of flour—in other words there will be no flour left. If this seems obvious, think how different it is from temperature measurements. You can have minus 5 degrees Fahrenheit or Celsius, but you can't have minus 5 pounds of flour.

In order to avoid the relativity of these two temperature scales, most scientists use an absolute temperature scale introduced by the British scientist Lord Kelvin near the end of the nineteenth century. On the Kelvin scale, zero degrees is the point where there is absolutely no heat. Zero degrees Kelvin is very cold—close to minus 460 degrees Fahrenheit—and is referred to as absolute zero. Nothing can get colder than absolute zero; in fact, it is a temperature that cannot actually be reached. On the Kelvin scale water freezes at 273 degrees, a comfortable room is about 293 degrees, and your body temperature is about 310 degrees.

Ahrens, C. Donald. *Meteorology Today: An Introduction to the Weather, Climate, and the Environment.* St. Paul: West Publishing Company, 1991.
"Fahrenheit, Gabriel Daniel," and "Thompson, Sir William." In *Dictionary of Scientific Biography,* Charles Coulston Gillispie, Editor-in-Chief. New York: Scribner, 1971.

The Fable of Centrifugal Force

If you tie a rock to the end of a piece of string and whirl the rock in a circle over your head, you feel tension in the string. Most of us would say that the tension comes from centrifugal force, a force pulling toward the outside of the circle, away from the center. But look more closely and you can see why centrifugal force is really a fable—a useful idea in some situations, but basically fictitious.

Imagine that while you're whirling the rock around, you let go of the string. If there really were such a thing as "centrifugal force" pulling the rock away from the center of the circle, that force would make the rock fly off in a direction straight away from the center of the circle. But the rock doesn't do that. Instead, it continues in the same direction it was going at the moment you released the string. The rock flies off along a tangent to the circle.

Obviously, whirling rocks on strings is dangerous unless you are alone in the middle of a very large open field. But you can safely observe how, according to the same principle, a baseball leaves a pitcher's hand along a tangent to the curved path of the hand. Both the rock and the baseball illustrate one of the laws of motion stated 300 years ago by Isaac Newton: Objects go in straight lines at steady speeds unless acted on by some force.

When you were holding on to the string, you were exerting a force on the rock, constantly pulling it toward you, keeping it from going in a straight line as it would, so to speak, prefer to do, and making it go in a circle instead. The tension in the string was caused not by "centrifugal force" but by you, pulling on the rock!

The same principle applies when you are in a car that makes a turn. Say you're steering to the right, and as you do you feel a mysterious force pushing you against the left side of the car, against the seat belt and the door. Souvenirs hanging from your rearview mirror swing to the left as if pushed by an invisible hand. Again, you might conclude that what we all call "centrifugal force" is pushing you away from the center of the turn. Centrifugal literally means flying away from the center. But why should a force appear out of nowhere just because you decided to turn right? What's really happening?

Consider another point of view—looking down on your car from a balloon hovering over the freeway interchange. You look down and see your car going along part of a circle as it rounds the corner. Remember Isaac Newton's law of motion. Some force must be acting on that car to make it go in a curve. The force in this case is friction between the road and the tires, pushing the car toward the center of the circle because the front tires are turned. Then, the car, its door, and its seat belts push the driver toward the center of the circle. If the friction were to disappear—say, because of ice on the road—the car would continue in a straight line, in accordance with Newton's law, in whatever direction it was moving at the moment it encountered the ice.

So in a tight right turn, the driver feels a push from the left side of the car and blames the sensation on "centrifugal force." But the big picture shows what's really happening: Friction between road and tires pushes on the car, and the car pushes on the driver, forcing them away from the straight path they would follow if there were no friction.

Abell, George O. "The Fable of 'Centrifugal Force.'" In *Exploration of the Universe*. New York: Holt, Rinehart and Winston, 1969.

The Storm Surge of a Hurricane

Many of us associate tornadoes and hurricanes with high winds and rain, but the most damaging force of a hurricane comes from the unusually high waves as the storm hits the coast. High winds create high waves, but in the case of a hurricane, the low pressure at its center also creates what meteorologists call a "storm surge."

When you suck water up through a straw, the water rises because the pressure in the straw is lower than the pressure above the water in the rest of the glass. Just like the suction that you create in the straw, the extremely low pressure at the eye of the hurricane can suck up the ocean into something like a hill of water. This is a "storm surge," and it can rise to as much as several yards above the water outside the stormy area.

The high winds that circle around the eye of the hurricane build up huge waves as much as 30 to 50 feet high that move off in all directions. As the hurricane reaches land, the storm surge, accompanied by the abnormally high waves, can flood areas far inland of where ocean water normally goes. When a hurricane coincides with a normal high tide, the water level—and the waves—are even higher.

In 1969, Hurricane Camille hit the coast of Mississippi. Camille's storm surge was more than 22 feet high. Combined with a normal high tide and winds of close to 200 miles per hour, the damage from Camille was estimated at $1.5 billion, and more than 200 people were killed.

Ahrens, C. Donald. *Meteorology Today: An Introduction to Weather, Climate, and the Environment.* St. Paul: West Publishing Company, 1991.

Savoring the Aroma

The nineteenth-century French chef Brillat-Savarin once wrote that "smell and taste form a single sense, of which the mouth is the laboratory and the nose is the chimney." Taste and smell are closely related in that they discriminate between different chemicals—unlike hearing and sight, which detect different frequencies of sound and light, or touch, which detects pressure or temperature.

When you smell coffee, receptor cells in your nose are correctly identifying molecules of coffee vapor drifting in the air. Taste buds are the receptor cells on your tongue that identify specific molecules in your mouth. Although taste and smell don't always detect the same chemicals, they work closely with each other and in similar ways. For a short distance, the mouth and nose share a common air passage called the pharynx. As the food passes through your mouth, vapors drift through the pharynx to your nose, so what your mind registers as the "flavor" of food is really a combination of its taste and its smell. When you exhale through your nose while eating, you get a stronger sense of what the food tastes like. And food tastes especially strong when it is heated because the warm food gives off more vapors.

When you have a cold, you can't taste your food as well for two reasons, both having to do with smell. When your nose is blocked, the vapors from the food don't drift through the pharynx so you don't get the smell. And some cold viruses can kill smell receptor cells so that even the vapors that do get through the pharynx won't have as much of an effect.

McGee, Harold. *On Food and Cooking: The Science and Lore of the Kitchen.* New York: Scribner, 1984.

Salting Your Food

Maybe you like salt on your food, but your doctor says it causes high blood pressure. Well, humans aren't the only animals with a taste for salt. Porcupines will chew through wooden outhouses for the salt left in the wood from human urine. Deer and other animals will search out natural salt deposits. A study by two psychologists in 1940 described a three-year-old boy whose damaged adrenal glands prevented his body from retaining sodium—the more important of the two elements in table salt. The boy recognized salt and ate it as another child might eat sugar. When he was put on a hospital diet with a normal amount of salt, the boy died. People, like all other animals, need the right balance of sodium in order to regulate nearly every system in the body. If they lose too much sodium, they crave the taste of salt.

And yet you can also develop a taste for higher or lower levels of salt in your food by using more or less of it. To figure out why *eating* more salt would cause you to *want* more salt, a team of scientists asked two groups of people to increase their salt intake. Members of one group added 10 grams of salt to their food each day, while members of the other group swallowed 10 grams of salt tablets. The group that added the salt to their food came to prefer highly salted food, while there was no change in the preferences of the group that took the tablets.

The conclusion the researchers drew was that among people with adequate salt levels in their bodies, the preference for salty food comes from acquiring a taste for salt, and is not a result of the increased sodium level in the body. So, although there is good biological reason for our enjoyment of salt, it appears we can still cut down on the amount of salt we eat and enjoy our food all the same.

Beauchamp, Gary K. "The Human Preference for Excess Salt." *American Scientist,* January–February 1987.
McGee, Harold. *On Food and Cooking: The Science and Lore of the Kitchen.* New York: Scribner, 1984.

Why Human Milk Is Low in Iron

A baby gets its first immunity from its mother. As the baby's own immune system gradually takes over, another change occurs to help fight off disease: the concentration of iron in the baby's body drops dramatically during the first few months.

All living cells need some iron, and so we normally think of iron as beneficial. But the bacteria that cause disease also depend on the iron in the baby's body. The baby has to maintain enough iron for its own body without developing an excess that could encourage disease-causing microorganisms.

So how does milk fit into this picture? Human milk helps starve the microorganisms of their iron in two ways. First, human milk is lower in iron than milk from most other mammals. Also, human milk contains a chemical called lactoferrin that is strongly attracted to iron. Lactoferrin binds up iron and prevents cells—including disease-causing bacteria—from using it. The relatively large quantity of lactoferrin and the low concentration of iron in human milk may be two of the reasons that breast-fed infants are less prone to infections than infants fed only on cow's milk or iron-fortified milk formula.

Anemia from lack of iron is a serious problem in many developing countries, but in more affluent societies, babies and adults may be getting so much iron that they are losing some of their natural immunity.

Kent, Susan; Eugene D. Weinberg; and Patricia Stuart-Macadam. "Dietary and Pro-phylactic Iron Supplements: Helpful or Harmful." *Human Nature* 1, no. 1 (1978).
Weinberg, Eugene D. "Iron Withholding: A Defense against Infection and Neoplasia." *Physiological Review* 64 (1984).

Floating

We've all heard that floating objects are held up by a so-called buoyant force. But it's a challenge to explain to yourself where this force comes from. You might think of it this way: Suppose you take a plastic bag whose volume is one cubic foot, fold it up and put it in your pocket, and jump in a lake. Underwater, you open the bag, fill it with water, tie it shut, and let go of it. The bag neither floats nor sinks. It neither floats nor sinks because it merely encloses a cubic foot of water that was already there in the lake to begin with.

But a cubic foot of water is heavy—about 64 pounds. Take that bag of water out of the lake, and it's hard to lift. What this experience tells you is that while that water-filled bag was im-mersed, the water in the lake was supporting it, pushing up on it with a force of 64 pounds.

Now here's the crucial point: The lake had, so to speak, no way of knowing what was in the bag. The lake water will push up with a force of 64 pounds on *any* immersed object whose volume is one cubic foot. Substitute a block of wood for the bag of water, and suppose you're underwater with a one-cubic-foot block of wood weighing 40 pounds. The wood's weight pushes it down with a force of 40 pounds, but the water pushes it up with a force of 64 pounds, because the block's volume is one cubic foot. The result: a net buoyant force of 24 pounds pushing the wood up toward the surface.

So the essence of this approach to understanding the subtle phenomenon of buoyant force is to imagine replacing a cubic foot of water with a cubic foot of something else. That something else will be supported as if it were a cubic foot of water!

Epstein, Lewis Carroll. *Thinking Physics.* San Francisco: Insight Press, 1989.

Radiation, a Word of Many Meanings

If someone says, "There's radiation in this area," you will probably stay out. The word "radiation," used in this sense, usually refers to what is technically known as ionizing radiation. Ionizing means taking electrons away from atoms or adding electrons to atoms. Where atoms are ionized, chemical changes are likely to take place. So ionizing radiation is radiation capable of causing chemical changes in material it strikes, including the cells of living organisms.

Ionizing radiation usually comes from decaying atomic nuclei. Decaying in this sense means spontaneously breaking into smaller pieces. A material made of atoms whose nuclei tend to decay is said to be radioactive. When the nucleus of an atom decays, most of the radiation that comes out is composed of some combination of three ingredients. First are the so-called alpha particles, each made of two protons and two neutrons stuck together. Second are beta particles, which are electrons. Third are gamma rays, which are high-energy photons—elementary particles of high-energy light. (The names alpha, beta, and gamma were given to the different components of nuclear radiation almost a hundred years ago, before anyone understood the particles involved.)

So when we hear about radiation as a dangerous phenomenon, we're usually hearing about ionizing radiation from the decay of radioactive nuclei. But the word "radiation" has many other uses in science and engineering, some with sinister connotations and some without.

In a wider sense, radiation means the propagation of energy through space. The light we see with our eyes is a form of electromagnetic radiation. The word "electromagnetic" is added because light is made of electric and magnetic fields vibrating from side to side as the light travels forward. Radio waves are another form of electromagnetic radiation.

You may hear a meteorologist talk about the radiation of heat from the ground to the sky. That process involves the type of electromagnetic radiation known as infrared rays. Infrared rays also give you a sensation of heat when you stand near a hot radiator. Microwave radiation is a form of electromagnetic radiation used to transmit telephone conversations and to cook food. Sound is described as radiation that carries energy through air and other materials, whereas electromagnetic radiation can travel through empty space. Recording engineers talk about the radiation of sound from a musical instrument or a loudspeaker. Using the word in an even broader sense, biologists may talk about the radiation of a species of living organism from one area into surrounding areas.

So the word "radiation" often refers to a dangerous emission of particles from decaying atomic nuclei. But radiation may refer to almost any kind of propagation of energy through space, or any process of divergence from a central point. Radiation is one of the most broadly defined and widely used words in science.

"Radiation." In *McGraw-Hill Encyclopedia of Science and Technology*, 6th ed. New York, 1987.

"Radiation." In *Oxford English Dictionary*, 2nd ed. New York: Oxford University Press, 1989.

40

Strict Rules for Sloppy Speech

Elocution is the study of the rules of proper speech. But what about the rules for sloppy speech? In fact, even sloppy speech has rules that native speakers know without ever being taught. Here's one example. When a four-year-old child asks, "Do I *hafta* go to bed?" you might call the word "hafta" a sloppy contraction of "have to." And yet by changing the *v* sound in "have" to an *f* sound in "hafta," the child is unconsciously following a complex rule of English that many non-native speakers never learn.

In speech, we have what are called "voiced" and "unvoiced" consonants. Hold your fingers against your throat so you can feel the vibration of your vocal cords. If you pronounce the sound of the letter *v*, you can feel your vocal cords working, and so *v* is what is called a voiced consonant. Next try the letter *f*. As long as you don't add any vowel sounds after the *f*, you can make that sound without the vibration, and so *f* is an unvoiced consonant.

Now, one rule of English is that, within a word, an unvoiced consonant—such as *f* or *t*—can't follow a voiced consonant, such as *v*. Or, conversely, a voiced consonant can't precede an unvoiced. That's why when a word ends in a voiced consonant, as in "bug," we pronounce the plural as if it were formed with a *z*-also a voiced consonant—instead of *s*, which is unvoiced.

It's not that we can't put an unvoiced consonant after a voiced; in a sentence such as "I have two cars," the *t* of "two" follows the *v* of "have" with no difficulty. It's only that within a single word, it isn't done. So when English speakers turn two words, such as "have to," into one word, "hafta," the *v* becomes an *f* so the word will seem natural to native speakers.

The Random Walk

The phrase "random walk" is a modern buzzword in book titles and party conversations. It actually refers to one of the most useful ideas in science. First, an example of a normal, non-random walk: You decide to walk to a point 10 steps in front of you. So you take 10 steps forward. That's a non-random walk.

Now, the simplest version of the random walk: Flip a coin. If the coin comes up heads, take one step forward. Tails, take one step backward. As you randomly step backward and forward, you gradually drift away from your starting point. There's no way to predict whether you'll end up ahead of or behind your starting point. But the theory of probability can predict about how many random steps it will take to carry you a certain distance either ahead of or behind your starting point.

If you want to get to a distance of 10 normal steps from your starting point, you'll probably have to take about 100 random steps. If you want to get 100 normal steps away, you'll have to take about 10,000 random steps—that's a lot more steps. In mathematical language, the number of random steps you'll need to take is roughly proportional to the square of the distance you want to cover.

This is scientifically important because your back-and-forth walk is a lot like the random motion of molecules and other small particles. Perfume molecules escaping from a bottle, salt molecules dissolving in water, and photons inside the sun all take random walks because they are constantly jostled by the particles around them. The theory of the random walk explains how fast the perfume smell travels, how long the salt takes to dissolve, and how long light takes to travel from the center of the sun to the surface.

Feynman, Richard P. "Probability." In *The Feynman Lectures on Physics*. Reading, Mass.: Addison-Wesley, 1963.

Gamow, George. "The Law of Disorder." In *One, Two, Three . . . Infinity*. New York: Viking Press, 1947.

Common Birthdays, Classic of Probability

Consider a class of 30 children. What is the probability that at least 2 of them have the same birthday? The surprising answer is that the probability is better than 70 percent that at least 2 children in a class of 30 have the same birthday.

The secret to understanding this amazing 70 percent figure is to think about the likelihood of all the children's birthdays being different. Imagine asking the children, one at a time, to announce their birthdays. The first child can have any one of 365 different birthdays, of course. The second child can have any one of 364 different birthdays that will not match the first child's birthday. In other words, the chance that the first two children's birthdays will not match is 364 out of 365.

Now the question becomes, what is the chance of getting 29 non-matches in a row? The third child can have any one of 363 different birthdays that won't match the first two. So the third child's chance of not matching is 363 out of 365. The fourth child's chance of not matching is 362 out of 365, and so on. With each new child, the chance of not matching birthdays with at least one of the previous children gets smaller and smaller. To find the probability of getting 29 non-matches in a row, you have to multiply all those chances together. A calculator makes it easy. And it turns out that the chance of getting 29 non-matching birthdays in a row is less than 30 percent. That's why the probability is better than 70 percent that at least 2 children in a class of 30 will indeed have the same birthday.

Gamow, George. "The Law of Disorder." In *One, Two, Three . . . Infinity.* New York: Viking Press, 1947.

Peters, William Stanley. *Counting for Something: Statistical Principles and Personalities.* New York: Springer-Verlag, 1987.

Note: The probability of a common birthday in a group of 30 is about 0.7304.

Babies on Treadmills

When your baby takes its first steps, it's on its way to mastering an act of balance and coordination so complex that scientists haven't yet been able to build a two-legged robot. But where does a baby, who couldn't even crawl a few months earlier, get such a complicated set of skills? In fact, babies are probably born with some of what it takes to walk.

To test babies' innate walking abilities, psychologist Esther Thelen at Indiana University held a group of infants so that their feet just touched a moving treadmill—something like a miniature conveyor belt. As the belt pulled their feet backward, the babies brought their feet forward one at a time. They could have brought their feet forward together or just let them drag behind, but for some reason the babies moved their legs as if they were walking.

Next, Thelen used a double treadmill with one belt under each foot. Even when the two belts traveled at different speeds, the babies kept on walking with regular, alternating steps, even though they had to move one foot much faster than the other to keep up.

Some babies walk more easily than others on the treadmill, but by around four months, most babies walk in a regular, alternating step. And seven-month-old babies walk on the treadmill as easily as older children walk by themselves. Babies begin walking around the age of one not because they learn how at that age, but because they have finally developed the strength, coordination, and balance to do what they have been partially capable of since birth.

A Ticklish Question

When someone else tickles the bottom of your foot, the nerves in your foot send messages to the brain which send you into convulsions of laughter. But tickle yourself and nothing happens. Very little is known about tickling—including why we laugh—but a group of British psychologists built a tickling machine to help explain why we can't tickle ourselves. They proposed two reasons why tickling ourselves doesn't work: first, we have control over the hand that is doing the tickling; and second, our brain gets information from our hand that changes the way it interprets the nerve impulses coming from the foot. This added information going to and from the brain somehow counteracts the tickle impulse from the foot.

To test their hypothesis, the psychologists used the tickling machine in three different ways. To operate the machine, someone stands on a box with a movable pointer sticking out of the top. By moving a handle on the side of the box, either the experimenter or the subject can drag the pointer around and tickle the foot.

When the experimenter operated the machine, the tickling worked. When the subject operated the machine, there was very little effect—just like tickling yourself. But when the experimenter moved the handle while the subject's hand rested passively on it, the level of tickling was in between self-tickling and being tickled by someone else. This method was still somewhat ticklish because the subject couldn't control the handle, but not as ticklish as the first method because the subject's hand still provided information about the movement of the pointer.

So, apparently tickling needs at least two other conditions to make us laugh: a lack of knowledge about what's going on, and a lack of control.

Bicycles, Footballs, and Space Shuttles

To steer a bicycle at high speed, you don't turn the handlebars, you just lean the way you want to go and the wheel turns slightly by itself. If you try to steer with the handlebars alone, you may fall over in the opposite direction.

The reason for this strange relationship between steering and leaning has to do with what physicists call "angular momentum." Any moving object—such as your bike—has momentum. Changing its momentum requires a force. You increase the momentum by pedaling; the wind decreases it by pushing in the opposite direction. But spinning objects—such as your bike's wheels—also have angular momentum. Like your forward momentum, the angular momentum of the wheel has a specific direction which is perpendicular to the ground as long as you're traveling upright in a straight line. By leaning the bike to one side or by turning the handlebars, you change the direction of the wheel's angular momentum, but the wheel reacts in a very surprising way. As you lean to one side, the wheel turns to the same side. If you steer to one side, the wheel wants to lean in the opposite direction.

Take the wheel from a bicycle and, holding it vertically with one hand on each end of the axle, give it a good spin. Tilt the wheel to the right as if you were leaning into a turn, and you'll find the wheel steers itself to the right. Turn the wheel as if you were steering, and you'll feel it lean the other way. The antics of the wheel are due to the changes in angular momentum you produce by twisting it in one direction.

The spinning bicycle wheel is very much like the device used to navigate the space shuttle—a gyroscope. A gyroscope is any object—but usually a wheel—spinning inside a stationary framework. The gyroscope's spin gives it angular momentum, as did the bicycle wheel's spin. Just as a moving object continues in the same direction because of its forward momentum, a gyroscope stays at the same angle because of its angular momentum.

Another way to feel the angular momentum of a gyroscope is to hold the base of a blender minus the bowl and blade in both hands and, without turning it on, tilt it to the right and left. Now tilt it with the motor on. You'll find that the spinning motor acts as a gyroscope with angular momentum and resists your efforts to change its angle. You may not feel a very strong resistance because the mass that is turning in this case is fairly small. But larger gyroscopes, mounted in the space shuttle, take the place of a compass for navigating in space. A navigational gyroscope sits in a framework that allows it to turn freely in any direction. When the space shuttle turns, the gyroscope's angular momentum causes the gyroscope to stay pointing in the same direction.

Navigators on land rely on the needle of a compass which always points in the same direction. In space, where there is no north, south, up, or down, astronauts use gyroscopes to tell which way they're heading.

When a quarterback throws a football, he spins the ball to give it angular momentum just like a gyroscope. The angular momentum stops the ball from wobbling and keeps it traveling point-first so as to offer the least wind resistance. Keeping the ball straight helps it go farther and also makes it easier to catch.

The Shape of a Raindrop

Ask a friend to draw you a picture of a raindrop and you'll probably get something the shape of a tear—heavy at the bottom with a long point on top. And if you look at a drop of water coming out of the faucet, it does have that shape. The water drop coming out of the faucet is shaped like that because of the attraction water molecules exert on each other. As the drop leaves the faucet, the last part to come out clings to the moisture inside the faucet, and so the drop is pulled out into a tear shape.

But the same attractive force of water molecules gives rain-drops an entirely different shape. Actually the shape of a raindrop depends on the size of the drop. Smaller raindrops are spherical because as the water molecules cling together, a sphere is the most compact shape. Larger raindrops, however, have another shape. Drops with a diameter of more than about two millimeters—or the diameter of the wire in a coat hanger—are shaped more like the top half of a hamburger bun. As the drop is falling downward, there's more air pressure on the bottom than on the sides. The mutual attraction between the water molecules still keeps the drop mostly round, but the air pressure on the bottom causes the drop to become flattened underneath.

If you're an artist, you may want to go on drawing raindrops in the shape of teardrops, since most people wouldn't recognize little circles or tiny hamburger buns as rain. But you might still want to know that, according to meteorologists, raindrops and tears have very different shapes.

Ahrens, C. Donald. *Meteorology Today: An Introduction to Weather, Climate, and the Environment.* St. Paul: West Publishing Company, 1991.

What Good Is Snow?

Any skier can tell you what good snow is, but when you're out shoveling your driveway in the winter, you may feel hard-pressed to find any good words for the white nuisance.

Actually, in cold climates, snow is far more useful than even most skiers might imagine. When the temperature drops below freezing, snow forms a blanket over the ground. Snow may be cold, but it's an excellent insulator against the colder air above. The snow protects plants and animals that can survive under the snow at temperatures slightly below freezing but would die in the outside air.

The winter snow blanket also keeps the ground from freezing too deeply. When a cold climate doesn't get enough snow and the ground freezes, spring rains and snow runoff from higher up can't seep into the ground. As a result, the plants on the hillside don't get enough water and the streams below become dangerous, raging torrents from the excess water.

Snow is a good insulator because the intricate crystalline flakes trap large amounts of air. Light, dry snow is the best because it contains the most air. Wet snow or old snow that is more compact is not as good an insulator. Human activities such as snowmobiling can also pack down the snow, reducing its insulation value and causing it to last later into the spring without melting.

A year without snow can affect warmer areas as well if they rely on water from rivers such as the Mississippi or the Colorado that start in northern mountains. Without spring floods from melting snow, lots of ordinarily active rivers are dry or low.

So if you live in a cold climate, try to remember this winter as you're out shoveling that you're working with nature's own quilt.

Ahrens, C. Donald. *Meteorology Today: An Introduction to Weather, Climate, and the Environment.* St. Paul: West Publishing Company, 1991.

A Fan of Sunbeams

The sun is setting behind broken clouds. The air is hazy. Sunlight shining through rifts in the clouds illuminates the haze and creates the appearance of a fan of sunbeams emanating from the sun. Even if both the sun and the clouds are below the horizon, out of sight, the effect is often visible. The fan-of-sunbeams effect can be produced not only by clouds but also by mountains on the horizon, blocking the sun's rays in some places and allowing them to pass in others.

After a moment's thought, you realize that the way those sunbeams appear to radiate in all directions from one point is really a perspective effect. It's true that the sunbeams all originate in one place—the sun—but the particular sunbeams you see lighting up the haze in the air do not actually radiate in all directions.

Because the sun is so far away—93 million miles—those

sunbeams are essentially parallel. They appear to radiate in all directions for the same reason that the rails of a straight railroad track appear to radiate from a vanishing point on the horizon. As the rails get farther and farther from you, they occupy a smaller and smaller portion of your entire field of view, so they appear closer and closer together.

Knowing this explanation for the fan-of-sunbeams effect, you can watch for related effects. For example, if the sunbeams are brilliant enough, you may see them not only diverging from the sun, but converging on the opposite side of the sky, just as railroad tracks appear to converge on the horizon behind you.

Finally, if you look down from an airplane at sunset, you may see sunbeams shining through rifts in the clouds. While people on the ground below you see those sunbeams appearing to radiate in all directions, you will see that they are actually parallel.

Humphreys, W. J. *Physics of the Air.* New York: Dover Publications, 1964.
Minnaert, Marcel. *The Nature of Light and Color in the Open Air.* New York: Dover Publications, 1954.

Making Water in the Desert

When Norwegian physiologist Knut Schmidt-Nielsen visited the Arizona desert, he was especially impressed by the number of small rodents. How, he asked, did these animals get enough water to survive? Kangaroo rats, like other desert rodents, rarely drink water, and they eat mainly dried seeds with almost no water. The answer, Schmidt-Nielsen found, lies in what's called "metabolic water," which is created when the body extracts energy from, or metabolizes, food.

If you hold a glass jar over a candle or the gas burner of a stove, you'll see moisture form on the inside of the glass. And yet there's no water in wax or natural gas. When the gas molecules burn, they separate into carbon and hydrogen. The hydrogen then joins with the oxygen in the air to form water, which condenses on the inside of the glass.

All organic matter, including the dried seeds of a kangaroo rat's dinner, includes carbon and hydrogen. When that food is broken down, the hydrogen from the food combines with oxygen that the animal has breathed. The result is metabolic water. Even though there's almost no water in the dried seeds, the body creates water when it metabolizes the seeds to get energy. A kangaroo rat gets about 90 percent of its water in this way, and the rest comes from the small amount of water already absorbed in its food.

Humans get about 10 percent of their water from metabolism, but our bodies are nowhere near as efficient as the body of the kangaroo rat when it comes to using water. In the desert, using water efficiently is a life-or-death matter, as is conserving water.

Conservation of water brings us to another interesting comparison between our bodies and the kangaroo rat's. When our noses drip in the winter, one reason is that the moisture from our warm breath condenses on the inside of our cold noses. The same process is at work when you breathe on a window pane on a cold day and mist appears on the glass from the moisture in your breath. But where does the moisture come from?

When you inhale a breath of air at 32 degrees Fahrenheit, it may contain as much water vapor as it can at that temperature. But as the air warms to body temperature, its ability to hold water increases tenfold. As a result, the warmed air immediately begins to absorb moisture from your lungs and throat. When your nose drips in the winter, one reason is that all that moisture in your breath is now condensing on the inside of your cold nose—just as it does when you breathe on a window pane.

But what does this have to do with desert rodents? Most people don't need to conserve body moisture because they can always drink water. And the fact that our bodies aren't adapted to conserve moisture suggests that humans evolved where water was plentiful. But in the desert a cold nose helps many animals conserve the moisture they need to survive.

Because the kangaroo rat's nasal passage is much longer, almost all of the water evaporated in the animal's lungs is condensed in its nose on the way out. That moisture is then reabsorbed instead of being exhaled into the dry desert air. Unfortunately for us, our noses don't reabsorb moisture. And that's why, unlike the kangaroo rat, we have to carry handkerchiefs to deal with a dripping nose.

Schmidt-Nielsen, Knut. *Desert Animals: Physiological Problems of Heat and Water.* London: Oxford University Press, 1964.
Schmidt-Nielsen, Knut. *How Animals Work.* London: Cambridge University Press, 1972.

Human on a Bicycle

A human on a bicycle is one of the most efficient machines on earth. In this case we are using the word "efficient" to mean the amount of energy needed to transport a kilogram of mass over a distance of one kilometer. Specialists in the field of biomechanics have found that in this sense a human on a bicycle is more efficient than a horse, a locust, a salmon, or any other running, flying, or swimming animal. A human on a bicycle is more efficient than a car or a jet airliner, although less efficient than a truck or a train. A human traveling one kilometer on a bicycle uses only about one-fifth as much energy as that same human traveling that same kilometer on foot, and goes about three times faster.

There are several reasons why bicycling consumes less energy than walking the same distance. In walking, our leg muscles must raise and lower our entire body with each step, as well as moving our legs repeatedly backward and forward. Walking also involves a lot of friction—with each step some of our energy goes into wearing out our shoes and socks.

On a bicycle only our legs go up and down, and the leg going up is lifted partly by the other leg pushing down on the other pedal. On a bicycle we sit at a constant height above the ground and do not need to put any effort into raising or lowering our entire body. Friction is kept to a minimum in bicycles through the use of ball bearings and pneumatic tires, both of which, incidentally, saw their first widespread use in bicycles a century ago, and were later incorporated into automobiles and other machines.

The main obstacle faced by a cyclist is wind resistance, which increases dramatically with increasing speed. But at moderate speeds a human on a bicycle is the most efficient moving creature we know.

Hunt, Robert G. "Bicycles in the Physics Lab." *The Physics Teacher,* March 1989.
Vogel, Steven. *Life's Devices: The Physical World of Animals and Plants.* Princeton, N.J.: Princeton University Press, 1988.
Wilson, S. S. "Bicycle Technology." *Scientific American,* March 1973.

Using Purple Cabbage as a pH Indicator

Maybe your shampoo advertises that it's "acid-balanced" or that it has a low pH. But how do you know? When chemists talk about the "pH" of a substance, they are referring to acidity and alkalinity on a pH scale ranging from 1 to 14. Anything above 7 is said to be alkaline. Anything below 7 is said to be acid. And 7 is the neutral point that isn't acid or alkaline.

One way to tell whether your shampoo is acid or alkaline is by using what's called pH paper. When you put a drop of shampoo on the paper, the paper changes to some new color depending on the pH. A chart usually comes with the pH paper indicating which colors refer to which pH value.

But if you don't have pH paper, you can use the broth left over from cooking purple cabbage. Straight out of the pot, the cabbage broth is slightly acid—a pH of about 6—and purple. If you slowly add acid, such as vinegar, the pH goes down and the color becomes red at a pH of around 3. If you add alkali, such as baking soda, the pH goes up and the color turns blue at a pH of around 8. A strong alkali, such as some scouring powder, will turn the broth green at around 11.

So, going from strong acid to strong alkali, purple cabbage broth goes through bright red, purple, blue, and green. The color changes look like magic tricks because the colors of the ingredients don't match the colors of the results, but what you're really seeing is the relationship between color and pH. Try using cabbage broth to measure the pH of other household chemicals, or you can even try using it to test the soil in your garden.

Tocci, Salvatore. *Chemistry around You: Experiments and Projects with Everyday Products.* New York: Arco Publishing, 1985.

Cooking with Alloys

Anyone who has used aluminum cookware knows that eventually the aluminum gets pitted and discolored. That happens as the aluminum reacts with the air and with the food that's cooked in it. And there's some possibility that aluminum absorbed from cookware may be harmful over a long period of time.

Many pots and pans as well as parts for boats are made of stainless steel because, unlike most metals, stainless steel won't rust. All steel is a combination—or "alloy"—of iron and other metals. Stainless steel doesn't rust because it contains about 15 percent chromium, which is much less reactive than iron. The chromium is also why stainless steel is often shiny, like a chrome bumper on a car.

But even with this benefit, stainless steel also has its drawbacks. One drawback to steel made with chromium—besides its high price—is that chromium doesn't conduct heat as well as iron. As a result, stainless steel pans tend to develop "hot spots" instead of spreading the heat out evenly like a cast iron pan. To make up for this problem, the best stainless steel pans have copper bottoms. Because copper is an excellent conductor, the heat spreads out more evenly.

So next time you're heating tomato sauce in a stainless steel pan, remember that the chromium alloy in the metal will keep the pan shiny, but may cause the sauce to burn.

McGee, Harold. *On Food and Cooking: The Science and Lore of the Kitchen.* New York: Scribner, 1984.

The Legacy of the Dodo

The last dodo bird died in the late 1600s, probably as food for sailors. On the island of Mauritius in the Indian Ocean, where they lived, the large, flightless dodoes had been shot and eaten for years by sailors. Today we think of the dodo as the classic example of a peculiar anachronism, but in its own habitat the dodo played a critical role.

Also on the island of Mauritius was the Calvaria Major tree, which had evolved seeds so hard that they couldn't germinate by themselves. Instead, the seeds had to be cracked somehow before they could grow into young trees. Dodoes ate these seeds and digested the outer layer. But by cracking the seeds and removing their outer layer, the dodo's gizzard also prepared the seeds for sprouting, and when they left the bird's body they were ready to grow. When the last dodo died, there were no animals left on the island that could perform the necessary service for the Calvaria seeds. Today there are only about a dozen of these trees left on the island, all more than 300 years old.

The relationship between the dodo and the Calvaria is one example of what biologists call "coevolution"—when two or more species evolve in ways that make them mutually dependent on each other. The dodo depended on the Calvaria for food, and the Calvaria depended on the dodo to make its seeds viable. Coevolution means that all species are part of a complex web, and that for every species that goes extinct, several more extinctions may result. Sometimes species adapt to life without their coevolved partners, but often, as in the case of the Calvaria, the dependent species die off as well.

Temple, Stanley A. "Plant-Animal Mutualism: Coevolution with Dodo Leads to Near Extinction of Plant." *Science*, August 26, 1977.

The Musical Bean

As more people look for substitutes for red meat, many of them are finding that beans, especially dried beans, don't digest very well. The result may be benign but disagreeable embarrassments at inopportune moments.

In recent decades, there's been an increase in research on flatulence, but the phenomenon itself is hardly new. In the Middle Ages, Saint Augustine saw flatulence as one more indication of man's fall from grace. For a long time it was believed that flatulence increased sexual appetite by tickling the genitals, and so Saint Jerome prohibited beans for the nuns under his charge.

In the 1950s, chemists identified the offending ingredient in beans as a group of chemicals called oligosaccharides, a term from Greek meaning "a few sugars." As the name implies, oligosaccharides are complex sugars made of a few simple sugars, but unlike some other complex sugars, they can't be broken down by the chemicals produced in the human digestive tract. As the undigested oligosaccharides move on to the lower intestines, bacteria do the work that our own bodies can't. In the process of breaking down the complex molecules, the bacteria produce a variety of gases, but mainly carbon dioxide—the same gas that we, and other animals, exhale all the time. In the 1960s, researchers estimated that the average adult produces about a pint of gas a day. Fortunately only a very small percentage of that has an offensive smell.

Not all beans contain equal amounts of oligosaccharides. Lima beans and navy beans are the worst offenders, and botanists are working on ways to grow beans with fewer oligosaccharides.

McGee, Harold. *On Food and Cooking: The Science and Lore of the Kitchen.* New York: Scribner, 1984.

Keeping a Cool Head

When we get hot, we sweat, and evaporation cools our body. When your dog gets hot, it pants. Air traveling rapidly through the dog's wet nose and across its tongue causes evaporation and cools its head. For a dog's body to get rid of excess heat, its blood has to carry the heat to the dog's head before the heat can be eliminated. So, while we cool down all over by sweating, a dog's cooling system works almost entirely from its head.

You might assume that cooling the whole body directly would be more efficient than cooling the head and letting the head cool the rest of the body. But in some ways, panting is more efficient than sweating. Warm-blooded animals maintain fairly constant body temperatures regardless of the temperature outside. But even warm-blooded animals get warmer from exercise, heat, or illness. Under these conditions, the increase in temperature may benefit the animal but not the brain. Even a small increase in the brain's temperature can cause serious damage.

How then do you let your body warm up while still keeping a cool head? One solution is in the way the head is separated from the body by a relatively thin neck. Just as the handle of a frying pan cools off faster than the rest of the pan, the head of an animal cools off faster than its body. The horns of some animals and the ears of the jackrabbit work like extensions on the frying pan handle and help get rid of heat even more efficiently. Like horns or jackrabbit ears, panting also helps get rid of head heat first. That doesn't mean that on a hot day your shaggy sheep dog is cooler than you are, but by panting, he's at least keeping a cool head.

Schmidt-Nielsen, Knut. *How Animals Work*. London: Cambridge University Press, 1972.

Benjamin Franklin Drops a Dollar

In 1757 Benjamin Franklin made some astute observations about heat and cold. Franklin noticed that if he touched the metal lock on his desk, then the wood of the desk, the lock felt colder, even though both were exposed to the same temperature in the room. He concluded that the metal lock felt colder because metal is a better conductor of heat than wood, and drew heat from his hand more than the wood did.

In 1757 Franklin could not have known the real nature of heat. Only in the last century has heat been revealed as random vibration of the atoms that make up all the objects around us. Faster vibration means hotter temperature. Those vibrations are conveyed from hot to cold objects when they touch. On a large scale, however, heat behaves very much like a fluid passed from object to object.

Benjamin Franklin's desk observations demonstrated that metal conducts heat faster than wood. Franklin demonstrated that fact even more dramatically with a silver dollar and a candle. He wrote: "If you take a dollar between your fingers with one hand, and a piece of wood, of the same dimensions, with the other, and bring both at the same time to the flame of a candle, you will find yourself obliged to drop the dollar before you drop the wood."

Franklin, Benjamin. Letter to John Lining, April 14, 1757. In *The Ingenious Dr. Franklin: Selected Scientific Letters of Benjamin Franklin,* ed. Nathan Goodman. Philadelphia: University of Pennsylvania Press, 1974.

From One Cell to Many

Some living things, like bacteria, are made of only one cell. Others, like human beings and trees, are made of many cells—maybe billions of them—each cell doing a special job, all cells cooperating. But some living things are right on the borderline between those two categories.

For instance, there's the whole order of green algae known as the volvocines. The microscope reveals that the basic cell of these algae is egg-shaped, with a couple of whiplike tails that it uses to swim around in water. The simplest volvocine algae are collections of just four of these cells, stuck together with a kind of jelly and arranged so their whiplike tails are all on the same side. The whole colony of four cells swims as a unit. There are other volvocines, with pretty names like Pandorina and Eudorina, made of 16, 32, or 64 cells that swim together as a unit, cooperating for the common good.

The next step up in complexity is to add specialization to cooperation. There are colonies of 64 or 128 cells—the so-called Pleodorina, for instance—in which only the cells on one side of the colony are capable of reproducing. The other cells help with swimming, but they can't reproduce.

And most complex of all are the so-called Volvox colonies, made of thousands of cells connected to make a beautiful emerald-green hollow sphere that swims as a unit. Inside each Volvox sphere are a few large cells whose only function is to reproduce. Those special reproductive cells aren't even capable of swimming.

The different kinds of volvocine algae, ranging from extreme simplicity to the beginnings of complexity, may be giving us a glimpse of how the first living things made of more than one cell evolved.

Curtis, Helena. *Biology*. 4th ed. New York: Worth Publishers, 1983.
Gilbert, Scott F. *Developmental Biology*. 4th ed. Sunderland, Mass.: Sinauer Associates, 1994.

Cologne and the Blue Sky

About a hundred years ago, the English physicist John Tyndall wrote an article about why the sky is blue. He wrote: "Eau-de-Cologne is prepared by dissolving aromatic gums or resins in alcohol. Dropped into liquid water, the scented liquid immediately produces a white cloudiness, due to the precipitation of the substances previously held in solution."

By "precipitation" Tyndall meant the formation of extremely small solid particles as the alcohol in the cologne mixed with the water.

Tyndall went on, "Against a dark background—black velvet, for example—the water . . . shows a distinctly blue color."

You can repeat Tyndall's experiment, using some of your own cologne. Put some tap water in a clear drinking glass, allow a few minutes for all the fine bubbles to rise to the top and disappear, then let a few drops of cologne fall into the water. The combination of cologne and water will make what looks like smoke—the precipitation Tyndall mentioned.

When you hold the glass up to a window, that smoky precipitate will look reddish in color. But when you view the glass against a dark background with light entering from one side, the same smoky stuff in the water will look blue, more or less like the sky.

Molecules of air act on sunlight in the same way that particles in the cologne mixture do. Extremely small objects like molecules tend to scatter blue light sideways while letting red light proceed straight through. Sunlight is a mixture of all the colors of the rainbow. Some of the blue part of sunlight is scattered every which way by air molecules, giving the clear sky its blue glow.

When the sun is low in the sky, its light passes through a lot of air on its way to us. Much of the blue is scattered out of sunlight by air along the way, leaving mostly reddish light for us to see.

Tyndall, John. "The Sky." *The Forum*, February 1888. Reprinted in *Fragments of Science*, 6th ed. New York: P. F. Collier, 1905.

Walker, Jearl. "The Colors Seen in the Sky Offer Lessons in Optical Scattering." *Scientific American*, January 1989.

Father Determines Sex

The microscope has revealed that every human cell contains 46 so-called chromosomes, which have been discovered in this century to be the carriers of genetic information. By genetic information we mean the instructions that determine superficial characteristics such as eye color and more basic characteristics such as sex. These 46 chromosomes come in 23 pairs. In each pair, one chromosome is inherited from the father and one from the mother. In every pair but one, the two chromosomes have basically the same structure and size.

One pair, however, may not match. This is the pair that determines sex. No matter what your gender, you have a so-called X chromosome, inherited from your mother, as one of the members of this pair. If you're male, you have a so-called Y chromosome inherited from your father accompanying that X from your mother. If you're female, the chromosome inherited from your father was an X, not a Y, giving you two X chromosomes in that last pair.

By the way, the word "chromosomes" literally means colored bodies. About a century ago, biologists found that things inside cells become visible through the microscope when the cells are stained with dyes. Chromosomes take certain dyes especially well, and appear as sausage-shaped objects in the cell.

So females have the combination XX in that sex-determining twenty-third chromosome pair; males have the combination XY. Your mother could have given you only an X chromosome, because human egg cells contain only X's; your father, on the other hand, could have given you either an X or a Y, because sperm cells can have either type. Your sex was ultimately determined by whether you received an X or a Y chromosome from your father.

Curtis, Helena. *Biology.* 4th ed. New York: Worth Publishers, 1983.

The Monarch Butterfly's Poison Pill

Milkweed plants defend themselves against being eaten by making a bitter-tasting poison of chemicals technically known as cardiac glycosides. Most insects and other animals who try to eat milkweed never try again, if they're lucky enough to survive the first attempt.

However, the larva of the monarch butterfly is special: it has evolved the ability to handle the toxins in milkweed, but not by digesting them and not by excreting them as waste products. Instead, the monarch butterfly larva stores up milkweed toxins in its body. Even after the larva has turned into a full-fledged orange-and-black adult monarch butterfly, its body still contains the milkweed toxins it ate early in life. Most creatures get sick or die after eating milkweed. But neither larva nor adult monarch suffers any harm.

Among the animals that like to eat butterflies are bluejays. However, a naive bluejay who tries eating a monarch butterfly may be in for the unpleasant surprise of a mouthful of bitter, nauseating milkweed toxin. The bluejay learns a lesson: don't eat orange-and-black butterflies. This is too late to help the individual butterfly attacked by the bluejay, of course, but it has obvious long-term benefits for monarch butterflies as a species.

So the monarch butterfly has evolved a sort of poison-pill defense. The butterfly uses the milkweed's defense system to make itself repulsive to its potential enemies. And if you're a butterfly living among hungry bluejays, it is to your advantage to be orange and black, to look like a monarch butterfly—even if you've never eaten milkweed in your life. You don't have to be toxic as long as you look toxic. That's the strategy of viceroy butterflies. They don't store milkweed toxin, but they look a lot like monarchs and are generally left alone by bluejays.

We've seen how this relationship with the milkweed plant benefits the monarch butterfly, and even butterflies that look like monarchs. But what's in this for the milkweed plant? After all, the monarch larvae do eat some of the plant. Why hasn't the milkweed evolved a toxin that repels monarch larvae as effectively as it repels most other animals? One possible answer is that monarch larvae concentrate the odor of the toxin by storing up the toxin and staying on the milkweed. That odor repels other potential enemies of the plant.

The story of the milkweed plant and the butterflies is fascinating. But there is no reason to think it's unusual. Among the plants and animals of the world there must be millions of other ingenious chemical defense strategies—most of them still unknown to us.

Harborne, J. B. *Introduction to Ecological Biochemistry.* 2nd ed. New York: Academic Press, 1982.

Cabbage Wars

Here's another true story about an insect, a plant, and a chemical made by the plant.

Cabbages and their relatives make an acrid-smelling mustard oil. This oil is the cabbage's way of defending itself. Actually the mustard oil is created in the cabbage from another chemical called sinigrin, which is toxic to most insects. Sinigrin will kill a black swallowtail butterfly larva. Most insects won't touch a plant containing sinigrin.

But experiments have shown that larvae of another butterfly—the so-called cabbage butterfly—won't eat plants that *don't* contain sinigrin. They'd rather starve. Adult female cabbage butterflies can be fooled into laying eggs on a piece of paper soaked in sinigrin. Sinigrin attracts cabbage butterflies and repels almost everything else. How did this arrangement come about?

It seems likely that cabbages, like most plants, evolved the ability to make toxins in order to repel animals that might eat them. But before long, in evolutionary terms, one particular insect, the cabbage butterfly, evolved the ability to eat the toxins without getting sick. That gave the insect a monopoly on the cabbage supply and freed it from having to compete for food with other insects. Now cabbage butterflies have evolved a taste for sinigrin. The aroma of the toxin serves as a signal that good food is nearby.

This leads to another question: Why haven't all the cabbages in the world been devoured by cabbage butterflies? Obviously it's in the interest of cabbage butterflies not to eliminate their entire food supply. But the individual butterfly doesn't know that.

Fortunately for the butterflies, there are other toxins and parasites that will cut down their population if it gets too large.

Harborne, J. B. *Introduction to Ecological Biochemistry.* 2nd ed. New York: Academic Press, 1982.

The Strangeness of Ice

Ice floats on water. We all know this and think nothing of it because water and ice are so common on the surface of the earth. Chemists tell us, however, that the buoyancy of ice is unusual.

Most liquids get denser when they freeze. This makes sense when you think about the nature of heat as revealed by twentieth-century physics. Heat is the tiny random jiggling motion of atoms. Faster motion means hotter temperature; slower motion means colder temperature. An atom in a cold object—that is, an atom that's jiggling slower—tends to take up less room than a fast-jiggling atom in a hot object. So most objects contract as they cool. And most substances weigh more per cubic inch when they're cold and solid than when they're hot and liquid.

Not so for ice. Water molecules can't link together to make an ice crystal unless the molecules are, so to speak, at arm's length from each other. Imagine everyone in a crowded elevator pushing everyone else out to arm's length and you get the idea. Water expands as it turns to ice. Ice weighs less per cubic inch than liquid water, so it floats.

Fish and other things living in water survive winter because ice floats. Lakes and oceans freeze from the top down. Ice on top protects liquid water below from the cold. Imagine the alternative: Suppose ice sank rather than floating. Ice would pile up on the bottom of every lake, where it would be protected from the sun and might never thaw. Soon every body of water would be permanently frozen, except for a thin liquid layer on top in the summer. Fortunately for life on earth, water is strange: when it's frozen solid, it's less dense than when it's liquid.

Atkins, P. W. "Molecules." In *Scientific American Library.* New York: Scientific American Library, 1987.

Animal or Plant?

Often the answer is neither. Before the twentieth century, animals included things that move, eat, and grow to a certain adult size and stop growing. Plants included things that don't move or eat and which grow indefinitely. Plants don't eat because they use the process of photosynthesis to make their own food from carbon dioxide, water, minerals, and sunlight.

Improvement in microscopes in the last century has undermined this simple scheme. Microscopes reveal that mushrooms, for instance, have a cell structure so distinctive that they can't rightly be called plants or animals. Some one-celled organisms make their own food by photosynthesis like plants, yet swim around like animals.

While the microscope was dissolving the old animal-versus-plant distinction, it was revealing another profound division among all living things. Some living cells have a nucleus, a separate body inside the cell that divides whenever the cell divides, and which contains most of the cell's genetic information. Other cells don't have a nucleus—the molecules with genetic information are distributed around the inside of the cell.

A cell with a nucleus is said to be eukaryotic; a cell without a nucleus is prokaryotic. We humans are eukaryotic—our cells have nuclei. So do the cells of all the other animals and plants we see every day. Bacteria are prokaryotic—they don't have nuclei. Animal or plant? That old distinction has been replaced. Now we might ask, prokaryote or eukaryote?

Many biologists favor a scheme of not two but five kingdoms of life. Here's a common everyday way of distinguishing between the kingdoms: think about a trip to the supermarket.

First stop: the dairy section. Obviously all these milk products ultimately come from animals.

Next: meats and seafood. Again, obviously animals.

Then produce. Celery, tomatoes, apples—all from plants. But what about mushrooms? Mushrooms grow out of the ground like plants, but there the resemblance ends. Mushrooms get nourishment from decaying matter on the forest floor, not from carbon dioxide and sunlight. Their cell walls are made of chitin, a substance that also makes up the outer armor of insects but which is not characteristic in plants. Mushrooms are neither plant nor animal—they are members of a distinct kingdom, the fungi. Other fungi at the supermarket include yeasts in the bakery and the molds in some cheeses.

Speaking of cheeses, the acids that help coagulate milk proteins to make cheese and yogurt usually come from bacteria, which are neither plants, nor animals, nor fungi—they are members of yet a fourth kingdom, the prokaryotes. Recall that prokaryotes are cells without a nucleus. The greatest value of this distinction has to do with history. Prokaryotes are relatively simple; that suggests that in the history of life on earth, prokaryotes came first.

Finally, it's over to the health and beauty shelf for a tube of toothpaste. The abrasive in that toothpaste is probably made from the mineral skeletons of diatoms, one-celled organisms completely different from bacteria. Diatoms live in the sea and are included in a fifth kingdom, the protozoa. Today's plants and animals probably evolved from protozoa of long ago.

Curtis, Helena. *Biology*. 4th ed. New York: Worth Publishers, 1983.

More Than an Ordinary Sauna

In 1775 a Dr. Blagden, then secretary of the British scientific organization known as the Royal Society, conducted a dramatic demonstration of the ability of the human body to maintain a constant internal temperature. Dr. Blagden and some friends stepped into a room that had been heated to a temperature of 260 degrees Fahrenheit. That's well above the 212-degree boiling temperature of water. *Don't try this.* They stayed in the 260-degree room for 45 minutes and emerged unharmed. A steak Dr. Blagden took with him, however, was cooked.

Unlike the steak, Dr. Blagden sweated profusely in the hot room, and that is why he was not cooked. Evaporation of water—sweat—takes energy—energy from heat brought to the skin by the bloodstream. Every molecule of water driven away from Dr. Blagden's damp skin carried heat with it, and that's why he survived in an environment far hotter than his body temperature.

Dr. Blagden demonstrated this principle in another way, with two buckets of water also taken into the hot room. One bucket had a layer of oil poured on top of the water, preventing evaporation of the water. The oil-covered water soon began to boil.

Another bucket contained water with no oil covering. That uncovered water stayed much cooler, because evaporation from the surface carried heat away from the water, just as evaporation of sweat from Dr. Blagden's skin carried heat away from his body.

You can see now that Dr. Blagden's survival in his 260-degree room depended on the air being dry. In fact, Dr. Blagden tried pouring water on the floor of the hot room to make the air humid. This reduced evaporation of sweat. Dr. Blagden had to get out of the hot room, fast.

Schmidt-Nielsen, Knut. *Desert Animals: Physiological Problems of Heat and Water.* London: Oxford University Press, 1964.

See Milk Protein

Put a teaspoonful of milk in the bottom of a glass. A teaspoonful is enough, because you're not going to want to drink this. Then add a teaspoonful of vinegar, wait a few seconds, and watch the milk curdle. The curd is mostly casein, the main protein in milk. What you've just done is not to change the chemical composition of the milk protein, but to make it behave differently by changing its environment.

Normally, small bundles of milk protein molecules are dispersed throughout the milk. They scatter light and contribute to milk's white color. When the environment becomes acid—when you add vinegar—electrical forces between proteins change. In an acid environment, milk proteins clump together and make the curd you see on the bottom of the glass after adding the vinegar. Tilt the glass to see the curd better.

Our demonstration mixture of milk and vinegar is basically useless, but curdling with acid under more carefully controlled conditions is one of the steps in changing milk into cheese. In cheesemaking the acid often comes from a carefully cultured strain of bacteria.

A completely different application of basically the same process is the mechanism that plugs small injuries in our skin. Proteins normally dispersed in our blood meet enzymes at the site of an injury. In the presence of those enzymes, the proteins link up into a web—a clot—that seals the wound and draws the injured skin back together.

McGee, Harold. *On Food and Cooking: The Science and Lore of the Kitchen.* New York: Scribner, 1984.

"Protein" and "Casein." In *McGraw-Hill Encyclopedia of Science and Technology,* 6th ed. New York, 1987.

66 *A Couple of Desk Tricks*

All you need for this experiment is a 25-cent piece and a small postage stamp. Put the stamp on the desk and hold the quarter horizontally about half an inch above the stamp. Now blow hard down onto the quarter. The stamp immediately rises from the desk and seems to stick to the back of the quarter until you stop blowing.

This is a quick and easy demonstration of the same principle that holds airplanes up in flight. A stream of air—or any fluid—has lower pressure than the atmosphere around it. The faster the stream, the lower the pressure. This is generally referred to as Bernoulli's principle, after the eighteenth-century Swiss physicist Daniel Bernoulli. The fast-moving air flowing past the edges of the quarter makes relatively low air pressure around the edge of the coin. Atmospheric pressure beneath the stamp then pushes the stamp against the quarter.

The wings of birds and airplanes exploit the same effect. The top surface of a wing is curved, the bottom relatively flat. As the wing moves through the air, that special shape forces air to move faster over the top of the wing than along the bottom. The faster-moving air on top of the wing has lower pressure than the slower-moving air on the bottom. The result is a force pushing up on the wing. If the wing is big enough and the air speed fast enough, that force can help carry a bird or an airplane through the air.

Here's another way to demonstrate the same principle. You need a candle, and a card such as a business card or playing card. Hold the card between your mouth and the candle, about two inches from the candle flame. Now blow against the center of the card. The candle flame will lean toward the card, not away from it, to the astonishment of all the guests.

In this candle trick, fast-moving air around the edges of the card makes low air pressure on the other side of the card. Atmospheric pressure on the far side of the candle pushes air toward the card, and the flame moves in response to that air.

Now pick up the wine bottle—or the soft-drink bottle or the ketchup bottle, depending on what kind of dinner it is—and hold it between you and the candle. Blow against the side of the bottle. The round shape of the bottle is streamlined—the shape makes it easy for air currents from your breath to rejoin on the other side of the bottle and continue toward the flame. The candle will be blown out as if the bottle weren't there at all.

Gardner, Martin. *Entertaining Science Experiments with Everyday Objects.* New York: Dover Publications, 1981.

Lynde, C. J. *Science Experiences with Home Equipment.* Princeton, N.J.: Van Nostrand, 1949.

Wine without Legs

Put some wine or some other alcoholic liquor in a glass, swirl the glass to wet the inside surfaces with liquid, set the glass down on the table, and wait. You are likely to see a ring of liquid clinging to the inside of the glass half an inch or so above the liquid surface. This liquid ring gets gradually thicker, until liquid begins to run down the side of the glass in streams sometimes called legs or tears.

More than 140 years ago, James Thomson, little-known brother of a famous English physicist named William Thomson, gave a lecture on this subject. His explanation involved surface tension—the attractive force between liquid molecules that pulls water into droplets on a well-waxed car, among other effects.

Thomson pointed out two important differences between alcohol and water: alcohol evaporates faster than water, and alcohol has less surface tension—weaker attraction between molecules—than water.

Thomson explained legs in a wine glass by saying that alcohol evaporates fastest from the liquid film on the glass. That leaves a watery film on the glass with less alcohol and stronger attraction between molecules than the wine below. Now the attractive force between water molecules pulls more wine up the side of the glass until the ring forms. When the ring gets too heavy, liquid runs out and makes legs.

A nice explanation, but can it be checked? In 1855 Thomson tested the idea by putting wine in a vial and sealing it with a cork to prevent evaporation of alcohol. No legs appeared. When the vial was uncorked and fresh air was allowed to reach the wine again, the legs reappeared. Wine in a corked vial has no legs. The experiment worked in 1855. Does it work in your kitchen?

McGee, Harold. *On Food and Cooking: The Science and Lore of the Kitchen.* New York: Scribner, 1984.

Thomson, James. "On Certain Curious Motions Observable on the Surfaces of Wine and Other Alcoholic Liquors." 1855. In *Popular Lectures and Addresses,* ed. William Thomson. New York: Macmillan and Company, 1891-94.

Unseen Reflection

In 1901 Guglielmo Marconi sent a radio signal across the Atlantic, from England to Newfoundland. How this happened was a great mystery to Marconi and everyone else. Radio signals were known to travel in straight lines, just as light rays do. Radio and light are really two forms of the same thing. So how did Marconi's signal get over the curve of the earth?

Two British engineers, Arthur Kennelly and Oliver Heaviside, independently proposed in about 1902 that there might be some unseen layer in the upper atmosphere that bends or reflects radio waves over the horizon. A British physicist, Edward Victor Appleton, looked for the so-called Kennelly-Heaviside layer, and found it in 1924. Appleton's technique was to send a signal from the English city of Bournemouth to Oxford. Appleton found that he could make the received signal stronger or weaker merely by changing the wavelength of the signal at the transmitter.

Part of the signal received at Oxford was coming straight from Bournemouth, and part was coming by way of the Kennelly-Heaviside layer—a longer path. If the two parts of the signal arrived at the receiver with their waves in step with each other, the received signal would be strong. If Appleton changed the wavelength, the received signal would be weak, because then a high point of one wave might arrive with a low point of another. The two waves, arriving out of step, would cancel each other out.

Edward Victor Appleton in 1924 not only showed the existence of the first known reflecting layer of what is now called the ionosphere, but calculated its height: about 50 miles above the ground.

By the way, this Appleton is no relation to the Victor Appleton who wrote the Tom Swift science-fiction novels. That Appleton was a pen name used by a syndicate.

"Appleton." In *Dictionary of Scientific Biography*, Charles Coulston Gillispie, Editor-in-Chief. New York: Scribner, 1971.
Appleton, E. V. "The Ionosphere." Nobel lecture, 1947.

Approaching the Dew Point

Take all the air in the lower atmosphere over the county you live in. That air contains a certain amount of water vapor. You can get an idea of how much water vapor by listening for the relative humidity measurement in a weather report. The relative humidity tells you how much water vapor the air holds, compared to the amount it's capable of holding. If the air over your county contains only 50 percent of the water vapor it's capable of holding, then the relative humidity is 50 percent.

The relative humidity will almost certainly change as the day goes on. More water vapor may be added to the air, possibly by evaporation from the ocean. But the biggest cause of change of relative humidity will probably be change of temperature.

This evening, as the air gets cooler, the relative humidity will increase. Cool air cannot hold as much water vapor as warm air. As the air cools, the amount of water vapor it already contains will get closer and closer to the maximum the air can hold. At some temperature the air will become saturated; the relative humidity will reach 100 percent; the air will become unable to hold any more water vapor. Fog may form. That temperature is the so-called dew point, another number sometimes included in weather reports. As air temperature approaches the dew point, relative humidity approaches 100 percent.

Any object whose temperature is at or cooler than the dew point will cause water to condense from the air. Water condenses on a glass of iced tea in summertime because the temperature of the glass is not only cooler than the air temperature, but cooler than the dew point.

During the night a blade of grass loses heat to the sky, like a radiator. A blade of grass can't hold much heat to begin with, so it quickly becomes cool enough to cause water to condense—in other words, its temperature drops below the dew point.

Ruffner, James A., and Frank E. Beri. *The Weather Almanac.* 5th ed. Detroit: Gale Research, 1987.

Note: For more on relative humidity, see "Humidity, Relative to What?" on p. 80.

Interpret Oil Stains

Imagine that your car has a very small engine oil leak. Sooner or later enough oil will leak out to make an oil drop hanging from the bottom of the engine like a water drop hanging from a leaky faucet. Soon the oil drop will be almost but not quite heavy enough to fall from the engine under its own weight.

Now suppose you drive the car through a dip in the road. As you climb the up side of the dip, you feel pressed down into your seat because the car is being accelerated upward by the slope. Meanwhile, the same upward acceleration throws that hanging oil drop from your engine. The oil drop falls to the concrete and makes a little stain.

Now imagine hundreds, thousands of cars with leaky engines

going through the same dip in the road. All those little oil stains add up to one big stain that says, "Here's a place in the road where cars accelerate upward." Of course, deep dips in the road are easy to see even without oil stains. But even a very gentle dip should, according to this line of reasoning, have a bigger, darker oil stain than a level part of the same road.

In fact, any part of a road where a car is made to accelerate upward or sideways is likely to have a bigger-than-average, darker-than-average oil stain. For instance, a car going down a hill will be accelerated upward when it encounters level road at the bottom of the hill. The driver will feel momentarily pressed down into the seat. And at the same time an oil drop may be thrown from the bottom of the engine onto the concrete. Oil drops may also be thrown from the engine when a car takes a curve at high speed.

So be an oil stain interpreter: see whether stains on highways near you reveal the shape of the road surface.

 71

Bartlett, A. A.; D. F. Kirwan; and J. Willis. "External Manifestation of the Variation of Free-Fall Acceleration inside Moving Cars." *The Physics Teacher,* November 1980.

You Can't Heat an Ice Cube

More precisely, you can't heat an ice cube past a temperature of 32 degrees Fahrenheit. Put some ice cubes in a saucepan, put the pan on the stove, turn on the heat. The ice will melt, but it won't get warmer than 32 degrees.

This strange fact troubled the eighteenth-century Scottish physicist Joseph Black. What really bothered him was, why do ice and snow melt so *slowly?* If you warm an ice cube to 32 degrees and then add just a little more heat, the ice doesn't melt all at once. You have to keep adding heat for a long time to melt the whole cube. Joseph Black concluded that adding heat does not always make things warmer. He wrote, "Melting ice receives heat very fast, but the only effect of this heat is to change it into water, which is not in the least sensibly warmer than the ice was before. . . . A great quantity . . . of the heat . . . which enters into the melting ice, produces no other effect but to give it fluidity . . . ;

[the heat] appears to be absorbed and concealed within the water, so as not to be discoverable by the application of a thermometer."

Joseph Black's observations of the 1750s made sense when the real nature of heat was revealed early in the twentieth century. Heat is the random jiggling motion of atoms; hotter temperature means faster motion.

Ice is water molecules linked up as a rigid crystal that can take a certain amount of molecular shaking without coming apart. But when the shaking is violent enough—when the temperature gets to 32 degrees—water molecules are shaken loose from the ice to form liquid water. As long as you continue to add heat to an ice cube, that heat is used exclusively to shake water molecules loose until no more ice is left. That takes time, so ice melts slowly—fortunately for people who live below snowy mountains.

Black, Joseph. "Lectures on the Elements of Chemistry." In William Francis Magie, *A Source Book in Physics.* New York: McGraw-Hill Book Company, 1935.

Wolf, Abraham. *A History of Science, Technology, and Philosophy in the Eighteenth Century.* 2nd ed. Gloucester, Mass.: Peter Smith, 1968.

Howling Wind

When wind blows over pine needles, the needles break up the smooth flow of air. Behind each pine needle is a wake in the air, something like the swirling water trailing a canoe paddle. This wake in the air is made of eddies—technically, vortices—that form just behind the pine needle, then detach themselves from it and move downwind.

These vortices form on alternate sides of the pine needle dozens or hundreds of times per second in a regular rhythm that creates a sound with a definite musical pitch. The pitch becomes higher if the wind blows harder—that is, if the air moves faster. The pitch is also higher if the diameter of the pine needles is smaller.

In the early 1900s, the Hungarian physicist Theodore von Karman mathematically analyzed the vortices that form behind cylindrical obstructions. It was a tough job, but von Karman developed a simple formula that allows you to predict what the pitch will be—in other words, how fast the vortices will form—if you know the diameter of the cylinder and the speed of the wind.

As the vortices, the eddies, move downwind from a cylindrical object, they take up alternating positions along the wake like lamps on opposite sides of a street. The vortices are now usually said to form a "von Karman vortex street" behind a cylindrical object. Decades ago, people often noticed a humming sound coming from telegraph wires strung between poles. The sound was caused not by electricity but by vortices forming in regular rhythm in the air downwind of each wire. Nowadays there are fewer aboveground wires to listen to, but we can still hear the weird sound of the von Karman vortex street when wind blows through a pine tree.

Humphreys, W. J. *Physics of the Air.* 3rd ed. New York: Dover Publications, 1964.
"Von Karman." In *Dictionary of Scientific Biography,* Charles Coulston Gillispie, Editor-in-Chief. New York: Scribner, 1971.

Measuring Altitude with a Thermometer

It takes longer to hard-boil an egg in Denver than in New York. The reason is that water boils at about 203 degrees Fahrenheit in Denver, and at about 212 degrees in New York. At Denver's altitude of about 5,000 feet above sea level, the atmospheric pressure is lower—fewer air molecules strike the surface of liquid water than at New York's altitude near sea level. So a pan of boiling water in Denver is slightly cooler than a pan of boiling water in New York because the altitude is higher, and the atmospheric pressure is slightly less. The boiling point of water drops about 2 degrees Fahrenheit with every thousand feet of altitude.

Once you heat water to its boiling point, you can't make the liquid any hotter. New heat energy added to boiling water goes into making water evaporate as steam, not into heating the liquid. Even if you turn up the gas all the way under a saucepan of boiling water in Denver, the water temperature will remain at 203 degrees—unless you seal the top of the pan so the pressure inside will increase. Then you have a pressure cooker.

If boiling temperature depends on atmospheric pressure, then why not measure atmospheric pressure by measuring boiling temperature? This clever idea occurred to Gabriel Daniel Fahrenheit himself in the 1720s. Fahrenheit marked one of his thermometers not with degrees but with a scale indicating atmospheric pressure. Just dip the instrument in boiling water, and you could read off the atmospheric pressure.

A Colombian geographer, Francisco José de Caldas, took the idea a step further around the year 1800. Knowing that atmospheric pressure decreases as you go higher, Caldas used the boiling point of water to measure the height of mountains in South America.

McGee, Harold. *On Food and Cooking: The Science and Lore of the Kitchen.* New York: Scribner, 1984.

Negret, Juan P. "Boiling Water and the Height of Mountains." *The Physics Teacher,* May 1986.

Rombauer, Irma S., and Marion Rombauer Becker. *The Joy of Cooking.* Indianapolis: Bobbs-Merrill, 1975.

Wolf, Abraham. *A History of Science, Technology, and Philosophy in the Eighteenth Century.* 2nd ed. Gloucester, Mass.: Peter Smith, 1968.

Relying on Bacteria

All protein molecules include nitrogen atoms. All living organisms need proteins, so they all need nitrogen. The air around us is 78 percent nitrogen gas, but our bodies cannot use nitrogen in that form. We must get nitrogen for our proteins either by eating plants or by eating animals that have eaten plants. Plants, in turn, get their nitrogen—some of it, at least—from so-called nitrates, nitrogen compounds dissolved in water in the soil. Those nitrates are made by bacteria when they decompose dead plant and animal tissue. So there is a nitrogen cycle: animals and plants die; their tissues are decomposed by bacteria to make nitrates; those nitrates are taken up from the soil by plants, which are eaten by animals, and so on.

But that's not all. Some soil bacteria break down nitrates and release the nitrogen to the atmosphere. These bacteria take usable nitrogen out of the biological cycle. If this lost nitrogen were not continually replenished, life on earth would soon disappear.

Fortunately, most of the lost nitrogen is replenished by so-called nitrogen-fixing bacteria. These bacteria provide a vital service: they capture nitrogen from the air and incorporate it into molecules plants can use—but not just any plants. Nitrogen-fixing bacteria live in the roots of the plants known as legumes—beans and alfalfa, for example. Farmers alternate corn and beans in the same field partly because nitrogen-fixing bacteria in the roots of the beans return nitrogen to the soil.

Nowadays usable nitrogen can also be added to soil in the form of chemical fertilizers. But most nitrogen fixation is still done by bacteria. Every living thing that relies on plants for food ultimately relies on nitrogen-fixing bacteria.

Curtis, Helena. *Biology.* 4th ed. New York: Worth Publishers, 1983.
"Nitrogen Cycle." In *McGraw-Hill Encyclopedia of Science and Technology*, 6th ed. New York, 1987.

Invention of the Vacuum Cleaner

According to the imposing eight-volume *History of Technology* edited by Charles Singer and Trevor Williams, an English civil engineer, H. Cecil Booth, got the idea for the vacuum cleaner about the year 1900. Booth went to a London hotel to see a demonstration of a new machine designed to clean the seats of railway cars. This American machine blew compressed air onto the seats to drive out dust.

Booth had the idea that this seat-cleaning device might work better if it sucked rather than blew air. He went back to his office, put a handkerchief on the carpet, knelt down on the floor, and sucked through the handkerchief. Sure enough, the other side of the handkerchief was covered with dust. Inspired by this experiment, Booth built a device in which an electric pump sucked air through a cloth bag, and the vacuum cleaner was born.

Soon H. Cecil Booth had created the British Vacuum Cleaner Company. The company's horse-drawn cart would visit people's homes, where uniformed workers would pass long suction tubes through windows into rooms to be cleaned. One of the company's most prestigious early jobs was cleaning Westminster Abbey for the coronation of King Edward VII in 1902. In 1904 Booth's company came out with a small domestic vacuum cleaner in which the electric air pump and cloth bag were mounted on a little wagon. The operator swept the carpet with a long suction hose plugged into the wagon.

In 1908 an American, James Spangler, developed a new compact design: an electric suction pump mounted vertically over a set of wheels. The pump sucked air and dust from the carpet and blew it into a cloth bag, rather than sucking through the dust bag as in Booth's version. Spangler sold the patent to the Hoover Company, which got out of its original business—leather processing—to make vacuum cleaners full-time.

Singer, Charles Joseph. "The Twentieth Century, c. 1900 to c. 1950, Part II." In *A History of Technology*, vol. 7, ed. Trevor I. Williams. Oxford: Clarendon Press, 1978.

A Stubborn Business Card

Take an ordinary business card—a calling card—and fold the two ends down at right angles, so the card makes a little bridge about half an inch high. Put the card on a table so it stands on the two folded edges. Now challenge someone to turn the card over by blowing air through the space underneath it. The card will not flip over. Instead it will cling to the table. The harder you blow, the tighter it clings.

This is another desk trick based on the principle that a fast-moving stream of air has lower pressure than the still air around it—a principle named after the eighteenth-century physicist Daniel Bernoulli, who discovered it. The fast-moving air under the folded card has lower pressure than the still air above. Atmospheric pressure above the card presses it onto the table.

Bernoulli's principle is the basis of flight. Airplane wings, like bird wings, are curved on top and relatively flat on the bottom. That forces air to flow faster over the top of the wing than across the bottom. That, in turn, makes lower air pressure on the top of the wing than on the bottom. The result is a force due to atmospheric pressure pushing up on the bottom of the wing.

Bernoulli's principle suggests that you might be able to flip the card over by blowing hard horizontally over the top of the card, so the fast-moving low-pressure air is above the card rather than below. It can be done, but it takes practice and luck.

Lynde, C. J. *Science Experiences with Home Equipment.* Princeton, N.J.: Van Nostrand, 1949.
Swezey, Kenneth. *Science Magic.* New York: McGraw-Hill, 1952.

Bent Out of Shape

Bend a paper clip slightly, and it springs back. Bend it too far and it stays bent, although it still has its springiness. These facts may not seem remarkable until you try to explain them. The French physicist Charles-Augustin Coulomb pondered the behavior of springy metals more than 200 years ago. Coulomb hung a weight at the end of a long metal wire, twisted the weight around the axis of the wire, and let go. If he twisted the weight only slightly, it would rotate back and forth, like a rotating clock pendulum or a rope climber spinning at the end of a rope, and eventually stop in its original position. But if he twisted the weight too far, it would eventually stop not in its original position, but in some new position.

Charles-Augustin Coulomb concluded that when he twisted the weight hanging at the end of the wire, he was working against two different kinds of resistance. First there was elasticity, or springiness, which returned the weight to its original position after a slight twist. Second, there was cohesion, which held the metal itself together and maintained the basic shape of the wire. Coulomb guessed that when the wire took on a permanent twist—that is, when it had been bent out of shape—it was because small component parts of the metal in the wire had slid over one another. The force of cohesion had been overcome.

That was more than 200 years ago. Much later, when the atomic structure of metals was revealed, Coulomb's hunch turned out to be right. When the atoms that make up a metal are pulled apart slightly, forces between atoms will pull them back toward their original positions. But if the force on the metal is great enough, whole layers of atoms will slide past each other. You can never get those layers back to their original arrangement. The metal is permanently bent out of shape.

Bronowski, Jacob. "The Hidden Structure." In *The Ascent of Man*. Boston: Little, Brown, 1974.

Wolf, Abraham. "Coulomb's Theory of Torsion." In *A History of Science, Technology, and Philosophy in the Eighteenth Century*, 2nd ed., revised by D. McKie. Gloucester, Mass.: Peter Smith, 1968.

Diving Raisins

Fill a drinking glass with some fresh carbonated beverage—any kind will do. Wait for the foam to settle. Now toss in a few raisins and wait. At first the raisins go straight to the bottom. Soon, however, they rise to the surface. After floating just beneath the liquid surface for a few seconds, each raisin appears to dive suddenly back to the bottom. Before long, each raisin rises again, dives again, and so on.

The raisins are actually carrying carbon dioxide out of the liquid in the form of bubbles stuck to their wrinkled skins. As the carbonated drink sits in the glass, it slowly loses carbon dioxide to the air anyway—that's why the drink goes flat after a couple of hours. Bubbles of carbon dioxide forming below the surface and rising to the top simply speed up the process.

But bubbles won't form spontaneously in the middle of a clean, quiet liquid. Bubbles form where the liquid touches some solid object. That object might be a small particle of dust or the surface of a raisin. Once a bubble begins to form, it tends to grow. The carbon dioxide dissolved in the liquid will, so to speak, take advantage of the chance to leave the liquid and become gas. So the bubbles on a submerged raisin grow until they lift the raisin to the surface. When the raisin reaches the surface, the bubbles eventually pop, although surface tension in the liquid may keep them just under the surface for a while. When the raisin has lost enough bubbles, it loses its buoyancy and sinks.

You might try cutting the raisins into small pieces before tossing them into the carbonated drink. Smaller pieces are lighter and therefore easier for the bubbles to lift. Some books recommend doing the same trick with coffee grounds or small shirt buttons.

Gardner, Martin. *Entertaining Science Experiments with Everyday Objects.* New York: Dover Publications, 1981.

Lynde, C. J. *Science Experiences with Home Equipment.* Princeton, N.J.: Van Nostrand, 1949.

Water Is Hard to Heat

Water is by far the most abundant liquid on the surface of the earth. Also, anywhere from 50 to 95 percent of the weight of any living thing is water. Water is familiar, but it's chemically peculiar.

One of water's peculiarities is that it is, so to speak, reluctant to change temperature, compared to other materials. You have to burn almost twice as much gas in a stove to raise the temperature of an ounce of water one degree as to raise the temperature of an ounce of ethyl alcohol one degree, for example. In chemists' language, water has a high specific heat compared to almost every other material.

The explanation takes us down into the molecular world. Recall that heat is the random jiggling motion of molecules. Faster motion means hotter temperature. Water molecules attract each other very strongly compared to molecules of most other materials. When you add energy to water in the form of heat, a lot of that energy goes into breaking the strong attraction between molecules rather than into making those molecules move faster. The result is that you have to burn a lot of gas under a saucepan to get the temperature of water in the pan to change even a little.

This reluctance of water to change temperature means that living things in the oceans don't have to contend with rapid variations in temperature. The sea is slow to cool at night and slow to warm in the morning, slow to cool in autumn and slow to warm in spring. The oceans are heat reservoirs; they provide a relatively constant environment for marine life and exert a moderating influence on the weather.

Curtis, Helena. *Biology.* 4th ed. New York: Worth Publishers, 1983.

Humidity, Relative to What?

Relative humidity compares the amount of water vapor the air actually holds at the moment to the maximum amount of water vapor the air is capable of holding. If the relative humidity is 100 percent, the air cannot hold any more water vapor—the air is said to be saturated. Wet laundry will not dry in saturated air.

On the other hand, if the air is not saturated—if the relative humidity is less than 100 percent—the air contains less water vapor than it's capable of holding. If the relative humidity is, say, 50 percent, then the air contains only half the water vapor it's capable of holding; if it's 75 percent, then three-quarters, and so on. Wet laundry will dry if the relative humidity is less than 100 percent because the air, so to speak, has room for more water vapor.

How much water are we talking about? Here's an example. Take some warm, humid summer air: temperature 86 degrees Fahrenheit, relative humidity 100 percent. Take enough of that warm, humid air to fill a phone booth. (The volume of a phone booth is assumed to be two cubic meters.) The water in that phone booth full of saturated summer air weighs about two ounces.

For contrast, imagine now a sample of humid winter air: relative humidity still 100 percent, temperature now minus 4 degrees Fahrenheit. Again, take enough to fill a phone booth. The water in that sample of saturated winter air weighs only about a fifteenth of an ounce. Cold air has a much smaller capacity for water vapor than warm air does. Relative humidity is a measure of how close the air is to saturation. It takes only a little water vapor to saturate cold air; it takes a lot to saturate warm air.

We now know that warm air can hold a lot more moisture than cold air. So why, when you turn on your furnace in the winter, does the heated air often feel so dry? How dry the air *feels* is a measure of how quickly moisture is being evaporated from your skin. As the relative humidity drops—in other words, gets farther below the saturation point—the air feels drier because it is taking more moisture from your skin.

On a cold winter day, there is probably much less moisture in the outside air than on a typical summer day in the same area. But the relative humidity may still be near 100 percent because at the lower temperature the saturation point is much lower. When that cold, saturated air goes from below freezing outside to room temperature inside, it doesn't lose any of its moisture, but it feels much drier because it is now capable of absorbing more than three times as much moisture from your skin. If the relative humidity outside is 100 percent, the relative humidity inside will be less than one-third of that—or only about 30 percent.

On a cold day in Milwaukee, there's less moisture in the air than on a hot day in the Sahara Desert, even though the relative humidity is higher in the colder climate. If you heat the air in Milwaukee to room temperature without adding any moisture, the relative humidity in your house will drop to less than what it is in the Sahara Desert.

Ahrens, C. Donald. *Meteorology Today: An Introduction to the Weather, Climate, and the Environment.* St. Paul: West Publishing Company, 1991.
Encyclopedia Britannica. 16th ed. Chicago, 1991.
Handbook of Chemistry and Physics. Cleveland, Ohio: Chemical Rubber Company, 1988-89.

Colors on Metal

Metal pots and pans often have patches of rainbow color on their surfaces after they've been used for a while. The colors are made by thin films of metal oxides—chemicals formed in a reaction between metal and air at high temperature. The colors look like the colors in oil floating on a puddle, or the colors in a soap bubble. In fact, all those colors are caused by the same mechanism: the reflection of light from both surfaces of a thin film.

Wherever the metal oxide film is exactly thick enough to accommodate a whole number of waves of, say, red light, red light will be reflected to your eye, because red-light waves bouncing off the back side of the film will be exactly in step with red-light waves bouncing off the front of the film. The same goes for other colors. Wherever the film's thickness varies, so do the colors.

Variations in time and temperature of heating make variations in the color of the metal oxide film. Toolmakers and metallurgists examine those colors to judge the temperature of a piece of metal during the process of tempering.

Around the turn of the twentieth century, the American glassmaker Louis Comfort Tiffany developed a way to apply a thin coat of metal oxide to glass to make iridescent colors. Lusterware and other Tiffany glasses now command good prices on the antique market.

Most metal oxide films stick tightly to the surface and protect the metal beneath. That prevents further reaction between metal and oxygen in the air. One big exception to this rule is iron. Rust, which is made of iron oxides, flakes off and exposes fresh metal, which rusts, flakes off, and so on until the iron is gone.

Brill, Thomas B. "Why Objects Appear as They Do." *Journal of Chemical Education,* April 1980.
"Corrosion." In *McGraw-Hill Encyclopedia of Science and Technology.* New York, 1987.

Hall of Mirrors

Hold two rectangular flat mirrors so their edges touch and so they meet at a shallow angle, like the covers of an open book. Now slowly bring the mirrors together as if you were closing the book. Watch the reflections. At some point the reflection of the right-hand mirror will appear in the left-hand mirror. At about the same time, the reflection of the left-hand mirror will appear in the right-hand mirror. Keep closing the angle between the mirrors. Soon the left-hand mirror will show not only a reflection of the right-hand mirror, but also what is reflected in that right-hand mirror, that is, the left-hand mirror.

And so it goes. As the angle between mirrors gets smaller, you see more reflected mirrors. Just before the two mirrors meet, you can peek into the space between them and see many reflections—dozens, perhaps. The more distant-looking reflections are dark because the glass absorbs some light with each reflection.

Now separate the two mirrors and hold them parallel and

facing each other. Better yet, prop the mirrors up on a table so you don't have to struggle with holding them steady. You have made a miniature hall of mirrors. Peek over the top of one mirror and you will see into what looks like a tunnel. If the two real mirrors are 6 inches apart, each reflected mirror will appear 6 inches behind the last one. In general, the reflection of any object 6 inches in front of a mirror appears 6 inches behind that same mirror. If you can see 10 reflections in your miniature hall of mirrors, then the tenth reflection will appear 60 inches behind the mirror; in other words, that tunnel will appear 5 feet deep.

Now, here's another mirror demonstration you can try. Turn off the lights and hold a candle about an inch in front of an ordinary bathroom mirror. Don't let the glass get hot. You'll see several reflections of the candle flame, apparently lined up behind the mirror. It's the hall-of-mirrors effect again. One of the two mirrors is the reflecting surface on the *back* of the glass. The other mirror is inside the *front* surface of the glass, which reflects some light back into the glass rather than letting it emerge into the air.

The general rule is: Put any object a certain distance in front of a mirror, and the reflection will look like a copy of that object exactly the same distance behind the mirror. Put an object between two mirrors, and the image of that object in one mirror becomes the object reflected in the other mirror.

If you open up a toy kaleidoscope, you'll see two mirrors joined at one edge like the covers of an open book. When you look through the eyepiece of the kaleidoscope, you look into those mirrors at a glancing angle and see multiple reflections of scattered pieces of colored plastic at the other end. Each mirror shows you not only a reflection of the colored plastic, but also a reflection of the other mirror—which shows you a reflection of the plastic and of the first mirror, and so on.

For this last demonstration, get a third rectangular hand mirror, and set the three mirrors in a triangle, facing each other, edges touching, like walls of a miniature triangular room. Drop a penny into the space between the mirrors. Peek over the edge of one wall and you'll see dozens and dozens of pennies. The Paris

Exhibition of 1889 had a room like this, big enough for people to walk into. Two or three people could become a crowd!

Greenslade, Thomas B., Jr. "Multiple Images in Plane Mirrors." *The Physics Teacher,* January 1982.

Take Bets on a Leaky Milk Carton

(Note: Be sure to do this experiment where it won't hurt if things get wet.)

Get an empty half-gallon milk carton. Take a sharp pencil and poke three small round holes in the side of the carton, one above the other: one hole about a quarter of an inch up from the bottom of the carton, a second hole about two inches higher, and a third hole about two inches higher still. Put this milk carton on a flat surface, fill it with water, and watch the water squirt out of the three holes. Take bets on this: Which of the three streams will hit the surface farthest from the carton?

You might guess that water will squirt fastest from the bottom hole—and rightly so, because the pressure is greater the deeper you go into a container full of water. Dive to the bottom of a swimming pool and feel the increasing pressure on your ears. So, because the water pressure is greatest behind the bottom hole, the water will squirt fastest from that hole. Therefore, you might guess, the bottom stream will hit the surface farther from the punctured milk carton than either of the other two streams. Try it. You may be surprised to see the bottom stream land *closest* to the milk carton. Either the middle or the top stream will hit farthest away.

What's wrong? We considered horizontal velocity, but we failed to consider the *time* needed for the water to fall from the height of each hole. Yes, the water coming out of the bottom hole is moving faster, horizontally, than water from the upper holes. But the bottom stream has less time to fall before it hits the flat surface. During that time the water can travel only a short horizontal distance.

But there is a way to make the demonstration come out as you might think it should. Try this: Move that same leaky milk carton

84

to the edge of the table, so the three water streams hit the floor. Now the bottom stream wins; it hits the floor farther from the table than either of the other two.

The distances from the three holes to the surface of impact are now much more nearly the same. Now the difference in time of fall is small between the top hole and the bottom hole, but the difference in horizontal velocity is still large. Now difference in horizontal velocity is the crucial factor, and the demonstration comes out as you might expect—the bottom stream wins.

Grimvall, Göran. "Questionable Physics Tricks for Children." *The Physics Teacher*, September 1987.
Paldy, Lester G. "The Water Can Paradox." *The Physics Teacher*, September 1963.

Bacteria, Corn, Cars, and Terre Haute

In the early 1900s the Russian-born chemist Chaim Weizmann, working in England at the University of Manchester, developed a process in which a certain strain of bacteria fermented corn into a useful chemical and an apparently useless by-product. The useful chemical was butyl alcohol—not the kind you drink. Weizmann expected that it would be useful in making synthetic rubber. The apparently useless by-product was acetone, now known to most of us as nail-polish remover. But that came much later.

Back to the early 1900s. World War I began. The Allies needed acetone to make the explosive cordite. Now the Weizmann process was used for a new purpose. The U.S. government took over distilleries in Peoria, Illinois, and Terre Haute, Indiana, and turned them into acetone factories using the Weizmann process. Butyl alcohol was the useless by-product now.

World War I ended. There was no longer much need for acetone. But butyl alcohol had been piling up during the war, and people in the paint business came up with a use for it. Before 1920 the bottleneck in car manufacturing was the paint shop. Automobile varnishes took up to three weeks to dry. Paint experts knew that the so-called nitrocellulose lacquers would dry much faster, but they needed a cheap solvent, among other things. Butyl alcohol filled the bill. A private company bought the distilleries in Peoria and Terre Haute and went into the business of supplying solvent for lacquer for new cars of the twenties—using the same chemical process Chaim Weizmann had developed because of his interest in synthetic rubber.

If there's a moral to this story, it might be that you can't predict the ultimate results of basic research.

Gabriel, C. L. "Butanol Fermentation Process." *Industrial and Engineering Chemistry,* October 1928.

———. "Development of the Butyl-Acetonic Fermentation Industry." *Industrial and Engineering Chemistry,* November 1930.

Haynes, Williams. *American Chemical Industry.* New York: Van Nostrand, 1945.

Ihde, Aaron J. *The Development of Modern Chemistry.* New York: Harper and Row, 1964.

86 *Big Shadows*

Try this after a candlelight dinner: Blow out all the candles but one, then turn off all the other lights in the room. Look at the walls. The candle throws frighteningly huge shadows of you and your friends onto the walls. The bigger the room, the bigger the shadows.

Being careful not to get burned, hold your hand about two feet from the candle and notice the size of the shadow your hand casts on the wall. Now move your hand to a position one foot from the candle; notice how much larger the shadow becomes. There's a mathematical proportion in this situation. If your hand is five times closer to the candle than to the wall, then the shadow of your hand will be five times bigger than your hand itself. If your hand is ten times closer to the candle than to the wall, the shadow will be ten times as big as your hand. And so on.

Why doesn't an ordinary living-room electric lamp cast huge, dark shadows? The answer is that electric lamps and most electric light bulbs are designed to cast soft shadows, unlike a bare candle flame. The candle on the dining-room table is a small light source; the flame is usually less than an inch high. A frosted light bulb, on the other hand, is several inches high and gives off light from all over its surface—it's a bigger light source than a candle. A lampshade makes the effective size of the light source even bigger. The larger the light source, the more diffuse the shadow.

If you put your hand a foot away from a candle, you block a large amount of light that would otherwise reach the walls; put your hand a foot away from an electric lamp with a frosted bulb and a big shade, and you block much less light, so the shadow on the wall is much less noticeable.

Lynde, C. J. *Science Experiences with Ten-cent Store Equipment.* 2nd ed. Princeton, N.J.: Van Nostrand, 1950.

The "Weightlessness of Space"

Actually space is not the only place weightlessness can occur. You can see weightlessness anyplace, anytime. Toss a bunch of keys on a key ring into the air. Toss the keys so they don't tumble end over end. After the keys leave your hand, while they travel through the air, they seem to float around just like weightless objects we've seen in television pictures taken aboard orbiting spacecraft. You can hear the keys strike each other gently as they float around. While they travel through the air, the keys don't hang down from the key ring anymore. The keys are weightless.

If you are in orbit in a space shuttle, you don't have to throw your keys to see this effect; you only have to let go of them. You, the spacecraft, and the keys have already been thrown—at about 18,000 miles an hour—by the rocket engines that fired for a few minutes after lift-off. You, the spacecraft, and your keys are in a continuous free fall, just as the keys were after you tossed them into the air at home.

Unlike the keys at home, you won't hit the ground if you're in orbit. True, your path constantly curves toward the center of the earth, because the earth's gravity constantly pulls you toward the center of the earth. But you're also going sideways at 18,000 miles an hour—about 180 times as fast as the fastest pitch in a major-league baseball game. Since you're above the atmosphere—in other words, in space—there's practically no air to slow you down.

So your path curves downward very gently—as gently as the curved surface of the earth below you. You keep falling, but you never get closer to the ground. Your orbiting spaceship and your keys fall at exactly the same rate as you do. So you and your keys can float inside the ship without hitting the walls. You and your keys are weightless.

"Weightlessness." In *McGraw-Hill Encyclopedia of Science and Technology,* 6th ed. New York, 1987.

The Mystery of Proteins

In the late 1700s, chemists observed that many materials change from solid to liquid when they are heated. Ice is an example. But a few materials change from liquid to solid when they are heated. Milk and eggs, for instance—and blood. Beginning in the 1700s, chemists called these unusual materials albuminous substances—the name was borrowed from albumen, another name for egg white.

At first, albuminous substances seemed to come mostly from animals. The watery juice left behind after any animal tissue was cooked in a pressure cooker would congeal. Later, albuminous substances were found in wheat and beans. By the mid-1800s a Dutch chemist, Gerardus Mulder, suspected that albuminous substances were basic materials of all living things. The Swedish chemist Jöns Jacob Berzelius suggested calling them proteins, from Greek words meaning "primary substance."

Not till the mid-twentieth century did anyone discover why proteins congeal. A protein molecule, it turns out, is made up of hundreds of smaller molecules, the so-called amino acids, hooked up to make a chain. The chain is coiled into a very specific shape for each type of protein. Heating uncoils protein molecules so that they can link together in new ways and make a relatively rigid structure. That's what happens when you cook an egg. The protein molecules act something like little balls of sticky string unrolling and then sticking together in a new, open mesh.

Incidentally, finding out the exact shapes of protein molecules is one of the biggest problems in modern chemistry, and one of the most important. How a protein functions in a living organism depends on its shape.

Ihde, Aaron J. *The Development of Modern Chemistry.* New York: Harper and Row, 1964.
McCollum, Elmer V. *A History of Nutrition: The Sequence of Ideas in Nutrition Investigations.* Boston: Houghton Mifflin, 1957.
McPherson, Alexander. "Macromolecular Crystals." *Scientific American,* March 1989.

Who Can Drink Milk?

Milk contains a particular kind of sugar technically called lactose. A molecule of lactose is made up of two simpler sugar molecules—glucose and galactose—hooked together. In that hooked-together form, the sugar is useless to the human body.

Infants and some adults make an enzyme in their small intestines that cuts the lactose molecule apart. This enzyme has the technical name of lactase.

Once the milk sugar molecule has been separated into glucose and galactose, the body can use it. That's how milk sugar—lactose—is digested. People whose small intestines don't make that lactase enzyme can't digest milk sugar. The undigested lactose passes into the large intestine, where it draws water from body tissues. It also ferments. That produces gas. The result is diarrhea, a symptom of lactose intolerance.

Normal human infants make the lactase enzyme until they're two or three years old. They need it to digest their mothers' milk. After that, most people don't make lactase anymore, and can't drink milk because they can't digest the milk sugar, lactose; they are lactose-intolerant.

But people who have studied diets around the world tell us that lactose intolerance is normal—if by normal you mean the situation that applies to most people in the world. Most adults in the world cannot drink milk—at least not more than about a pint a day. Adults who can drink milk tend to be descended from cultures with a long-standing tradition of raising dairy cattle; their ancestors come either from northern Europe or from certain areas of Africa. Exactly how the genetic trait of adult tolerance for milk sugar became concentrated in those areas of the world is still unknown.

Kretchmer, Norman. "Lactose and Lactase." *Scientific American,* October 1972. Reprinted in the anthology *Nutrition,* with introductions by Norman Kretchmer and William van B. Robertson. San Francisco: W. H. Freeman, 1978.

McGee, Harold. *On Food and Cooking: The Science and Lore of the Kitchen.* New York: Scribner, 1984.

Reflections in the Water

Springtime: Magazines run vacation advertisements, illustrated with beautiful photographs of faraway places. Often these photographs show some landscape reflected in a quiet lake, with water smooth as glass. Maybe you've taken pictures like this yourself. The water is so smooth that the upside-down reflection is indistinguishable from the real scene.

Or is it? When you look at the reflection of a house in the water, you actually see the house from a slightly different point of view than when you look directly at the real house.

Imagine following your line of sight as you look at a reflection in a smooth lake. Draw an imaginary straight line to the point on the lake where you see the reflection of some particular feature on the house. Now draw another straight line from that point on the lake to the corresponding point on the real house. That second straight line goes *up* from the water to the house.

Those imaginary straight lines are the paths followed by light on its way from the house to your eye.

So when you look at the reflected image of a house, you are in effect looking *up* at the house. Any object in front of the house—a person standing in the front yard, for instance—will look a little taller, compared to the house, in the reflection than in reality, because that person is being viewed from a slightly lower vantage point.

So you may be able to distinguish the reflected scene from the real one. The reflected landscape shows everything from a lower point of view.

Easton, D. "On Reflections in Ponds." *The Physics Teacher*, March 1987.

Half Heads, Half Tails

Flip an honest coin, and you expect a head just as much as you expect a tail. Flip a coin 10 times, and you expect to get about half heads and half tails. But here's an interesting question: Should you expect to get exactly 5 heads and 5 tails?

The answer is no. It is very likely that 10 flips will give

approximately half heads and half tails. But it is much less likely that 10 flips will give *exactly* half heads and half tails.

Here are some more precise numbers, calculated from the laws of probability. If you flip a coin 10 times, there is less than a 25 percent chance of getting exactly 5 heads and 5 tails. However, there is almost a 66 percent chance that you'll be close to 5 heads and 5 tails; in other words, there's almost a 66 percent chance that you'll get either 4 heads and 6 tails, 5 heads and 5 tails, or 6 heads and 4 tails.

The more times you flip the coin, the more likely it is that you will get *approximately* half heads and half tails; but the more times you flip the coin, the less likely it is that you will get *exactly* half heads and half tails. If you flip a coin 20 times, your chance of getting exactly 10 heads is only about 18 percent, but your chance of getting something between 8 and 12 heads is about 74 percent.

If you flip a coin a million times, your chance of getting exactly 500,000 heads is infinitesimally small; but your chance of getting between 499,000 and 501,000 heads is extremely large.

Feynman, Richard P. "Probability." In *The Feynman Lectures on Physics*. Reading, Mass.: Addison-Wesley, 1963.
Huff, Darrell. *How to Lie with Statistics*. New York: Norton, 1954.

Note: The numbers come from the binomial distribution tables in the Handbook of Chemistry and Physics.

Spiders Don't Get Caught in Their Own Webs

Spiders have an oily secretion on their feet that keeps them from sticking to their webs. But there's more to the story than that. Not all the threads made by a spider are sticky. A spider can make more than one kind of silk. Spiders use their silk not only to make the sticky parts of their webs, but to line their burrows, to wrap their eggs, and as parachutes enabling them to travel on the wind.

Even in a web whose function is to trap insects, not all the threads are sticky. Take, for example, the so-called orb web—the kind that probably comes to mind first when we think of a spider web. An orb web has one set of threads emanating from the center, and another thread applied over those radial threads in a spiral pattern something like the groove on a phonograph record. Generally, only the spiral thread is sticky, not the radial threads.

But even the sticky threads in a web are not sticky over their entire length. A spider makes a sticky thread by applying glue to the silk as it is spun. The spider applies glue continuously to the new thread as it emerges from the end of the spider's abdomen. This glue doesn't form a continuous coat on the silk thread, however. Surface tension comes into play.

Surface tension is the same attraction between molecules that makes water gather into beads on a well-waxed car. Surface tension in the spider's glue gathers that glue into beads, arranged along the thread like beads on a string. Those beads of glue on the spiral threads of an orb web may be just barely visible through a good magnifying glass.

Bristowe, W. S. *The World of Spiders.* London: Collins, 1971.
Grzimek's Animal Life Encyclopedia. New York: Van Nostrand Reinhold Company, 1984.
Zim, H. S. *Spiders and Their Kin.* New York: Golden Press, 1990.

Ice in Oil

Fill a small drinking glass with vegetable oil at room temperature, add an ice cube, and watch what happens. The ice cube will probably float. But as the ice melts, water gathers in a large drop on the bottom of the floating ice cube. Soon this drop of water becomes heavy enough to separate from the ice cube and fall slowly to the bottom of the glass. The water drop sinks because water is denser than vegetable oil. The weight of a sphere of water half an inch in diameter is greater than the weight of a sphere of vegetable oil of the same size.

You can see liquid water coming off an ice cube in oil because oil and water don't mix. The reason for that has to do with a basic difference in the structure of oil molecules and water molecules. A molecule of water has a slight positive electrical charge on one side and a slight negative charge on the other side. The charge here is a very tiny amount of the same kind of charge that everybody knows as static electricity. In the world of electricity, opposite charges attract. So the positive side of one water molecule is strongly attracted to the negative sides of other water molecules; in general, water molecules attract each other strongly. Molecules of oil, on the other hand, don't have these electrical charges. So water molecules won't get into intimate contact with oil molecules.

Incidentally, this demonstration doesn't really show how fast ice melts in liquids other than oil, such as water. Speed of melting depends on other factors such as how water molecules from the ice interact with molecules in the surrounding liquid and how quickly the liquid conducts heat to the ice.

Webster, David. *More Brain-Boosters.* Garden City, N.Y.: Doubleday, 1975.

The Fuzzy Edge of a Shadow

Walk outside on a sunny day, hold your hand out in front of you, and look at the shadow of your hand on the sidewalk. The edges of the shadow are not sharp but fuzzy. The sun cannot cast a perfectly sharp shadow because the sun is not an infinitesimally tiny point of light. The sun appears as a disk, about one-half a degree of angle in diameter.

To see why a disk of light cannot cast a sharp shadow, imagine yourself as some kind of tiny bug walking on the sidewalk through the shadow of someone's hand, looking up toward the sun as you go. At first you walk in full sunlight. Soon, however, you enter a region where the hand overhead blocks out part of the sun's disk—but not all of it. In that region, you, the little bug, get something less than full sunlight. You are in the fuzzy gray edge of the shadow, the region technically called the penumbra.

You, as the little bug, continue walking. You enter regions where more and more of the sun's disk is blocked by the hand above. Eventually you progress from the penumbra to the so-called umbra, the dark central part of the shadow, the region in which the sun's disk is completely blocked by the hand above.

Keep going and you soon emerge into the penumbra on the other side. From there you eventually return to full sunlight.

Not all shadows have an umbra, a dark central part. Imagine yourself as a bug on the sidewalk again, but this time imagine the person whose hand is casting the shadow to be standing on a balcony, high above the sidewalk. From your bug's vantage point on the sidewalk, the hand appears smaller than the sun's disk, so you can never see the sun completely blocked by the hand. You are always illuminated by at least part of the sun's disk. The person whose hand is casting the shadow looks down at the sidewalk and sees only a penumbra, a fuzzy gray shadow of the hand.

Minnaert, Marcel. *The Nature of Light and Color in the Open Air.* New York: Dover Publications, 1954.

Brighter Colors from the Air?

When we look at a distant mountain, we see not only light reflected from the surface of the mountain but light added by the intervening air. The light added by the air is bluish, just like the blue of a clear sky. So distant mountains look blue.

The sun's light is a mixture of all the colors of the rainbow—it's white. But the portion of sunlight scattered sideways by air is bluish. It happens that very small particles like air molecules tend to scatter blue light sideways while allowing red light to pass straight through. So as a beam of sunlight passes through air, some of the blue is removed from the beam and sent out to the side. It is that blue light that we see when the sun shines down through the air between us and a distant mountain. The more air between us and the mountain, the more blue is added to what we see when we look at the mountain.

Now suppose we fly over this mountain in an airplane. If we are, say, 2 miles above the mountain, we look through much less air to see it than if we are on the ground, 50 miles from the mountain. Therefore, less blue is added to what we see; if the air is clean, the mountain is likely to look more vividly colored from the air than from the ground.

The book *The Nature of Light and Color in the Open Air* by Marcel Minnaert suggests that a landscape seen from an airplane at low altitude has more vivid colors than that same landscape seen from the ground. Professor Minnaert writes: "The haze hiding all colours as long as we were on firm ground, has practically disappeared altogether, and the hues are now displayed for the first time in their full warmth and saturation. This explains the charm of the scenery experienced by everyone who has had the opportunity of making a trip by air."

Minnaert, Marcel. *The Nature of Light and Color in the Open Air.* New York: Dover Publications, 1954.

Some of the complexity of atmospheric color is discussed by Jearl Walker in his "Amateur Scientist" column in the January 1989 *Scientific American.*

Find the Center of Your State

Once in a while we hear the name of a Midwestern town that is said to occupy the geographical center of the continental United States. What this statement usually means or implies is that if you go to that town, as many square miles of the continental U.S. will be north of you as south of you, and as many square miles of the continental U.S. will be east of you as west of you.

Where's the geographical center of your own state? Here's a way to get a good idea. Cut out a piece of cardboard in the shape of your state. Get the shape by tracing a good map. Now stick a pin perpendicularly through the cardboard somewhere near the edge. Tie a thread to this pin. Attach a small weight to the other end of the thread—a paper clip, for instance. Now hold the pin horizontally so the cardboard and the weighted thread hang straight down from the pin.

The so-called center of gravity of the vertically hanging cardboard—the point at which it will balance—is somewhere directly below the pin, somewhere along the line marked by the vertically hanging thread. So you want to draw a pencil line on the cardboard exactly where the thread is now. Hold the thread against the cardboard and draw a line exactly where the thread lies.

Now stick the pin through the cardboard in some other place near the edge, so that the thread crosses the line you have drawn. Again, let the cardboard and thread hang freely, then hold the thread against the cardboard and trace its position with the pencil.

The two lines you've drawn should intersect at the center of gravity, the point at which the cardboard will balance on the end of a pencil eraser. That point corresponds to the geographical center of the state. Now refer back to your map and see which town is closest to that point.

Gardner, Martin. *Entertaining Science Experiments with Everyday Objects.* New York: Dover Publications, 1981.

Goldstein-Jackson, Kevin. *Experiments with Everyday Objects.* Englewood Cliffs, N.J.: Prentice-Hall, 1978.

More Than One Way to Make a Frog

Most frogs in North America lay eggs in water. Then the eggs hatch into tadpoles, which have tails for swimming and gills to gather oxygen from the water. Eventually the tail and the gills disappear as the tadpole develops into an adult frog. But some frog species, especially in Central and South America, have a different life history: they skip the tadpole stage. In many tropical species, the eggs are laid in burrows in the ground, not in the water. The newly hatched frogs already look something like miniature adults. They never develop gills or even the circulatory system that would supply blood to gills. This so-called direct development, without a tadpole stage, seems to be well suited to tropical environments, where tadpoles swimming in open water would be in greater danger of being eaten than in colder climates like those in North America.

So different types of frogs have different ways of developing from embryos into adults. The difference in life histories is due not so much to a difference in basic body plan but to a difference in timing of development. For example, one type of tropical frog that skips the tadpole stage begins to develop legs very early— long before it hatches. In North American frogs, on the other hand, the development of legs is delayed until long after hatching. So tadpoles have no legs. But the end results—the legs of the adult frog—are basically the same in tropical frogs and in North American frogs. This is a case in which the schedule of development from egg to adult has been modified by evolution to suit different environments.

Raff, Rudolf A., and Thomas C. Kaufman. *Embryos, Genes, and Evolution.* New York: Macmillan, 1983.

Dappled Shade

As new leaves grow on the trees, the forest floor becomes dark in the middle of the day—except for small circular patches of sunlight here and there on the ground. Those patches of light are circular because they are images of the circular disk of the sun itself, projected on the ground by small openings between the leaves overhead. The principle of a pinhole camera is involved.

Since light travels in straight lines, a small hole can project an image on a flat surface. Each point in the projected image of the sun is formed by a ray of light coming through the opening in the leaves from one point on the disk of the real sun.

Although you can't usually tell, the images of the sun cast on the forest floor are upside down. Light from the north side of the sun passes through the opening in the leaves and ends up on the south side of the projected image; light from the south side of the sun ends up on the north side of the image. More than a thousand years ago, a Chinese philosopher, Shen Kua, explained this by pointing out that light rays going through a small hole are constrained like an oar in a rowlock: when the handle is down, the paddle is up, and vice versa.

Occasionally you may get a chance to verify that those little circles of sunlight really are upside-down images of the sun. If a cloud drifts across the face of the sun, moving, say, from west to east, then the shadow of that cloud in each little circle of light on the ground will move the opposite direction—from east to west.

The sun appears as a half-circular disk when it is either half risen or half set over the ocean or behind any other smooth, flat distant horizon. Suppose that, during such a sunrise or sunset, the sun's light shines through a tree onto a smooth wall. The patches of light on the wall will be not circles but half-circles, with the flat side up—upside-down images of the sun's half-disk.

Minnaert, Marcel. *The Nature of Light and Color in the Open Air.* New York: Dover Publications, 1954.

The Chinese analogy comes from J. H. Hammond, *The Camera Obscura: A Chronicle.* Bristol: Adam Hilger Ltd., 1981.

The First Elementary Particle

In the 1890s, physicists all over Europe were experimenting with electric currents flowing between two metal plates inside a sealed glass bulb with almost all the air pumped out.

In one variation on this scheme, electricity emanated from a metal plate connected to the negative terminal of a battery, passed through a metal tube connected to the positive terminal of the battery, and hit the other end of the glass bulb, where it made a glowing green spot. A TV picture tube is basically a fancy version of the same gadget. But in the 1890s people wondered: What was that beam of electricity in the glass bulb made of?

The English physicist Joseph John Thomson investigated the beam by carefully measuring how much it was deflected by static electricity and by a magnet. Thomson's method was something like watching someone throw ping-pong balls and golf balls in a high wind. Even from a distance, you can distinguish ping-pong balls from golf balls, because ping-pong balls, being light in weight, are deflected by wind much more than golf balls. Watch the deflection, and you can distinguish light balls from heavy ones.

By watching how much the mysterious beam in the glass bulb was deflected not by wind but by electric and magnetic forces, J. J. Thomson concluded, after some calculation, that the beam was made of particles almost 2,000 times lighter than the lightest atom. Today those particles are called electrons. All atoms, of every kind, contain electrons.

In 1897 J. J. Thomson concluded that he had found "matter in a new state, . . . in which the subdivision of matter is carried very much further than in the ordinary gaseous state: a state in which all matter . . . is of one and the same kind; this matter being the substance from which all the chemical elements are built up." Today J. J. Thomson is remembered for discovering the first elementary particle, the electron, in 1897.

Thomson, Joseph John. "Cathode Rays." 1897. Reprinted in *The World of the Atom*, ed. H. Boorse and L. Motz. New York: Basic Books, 1966.
Toulmin, Stephen E., and June Goodfield. *The Architecture of Matter*. New York: Harper and Row, 1962.

Pillbugs Live in Damp Places

Pillbugs, woodlice, and sowbugs live in damp places because they have gills, and gills won't function unless they are wet. Pillbugs, woodlice, and sowbugs are all members of a suborder—technically called the land isopods—within the class of crustaceans. Other familiar crustaceans include lobsters, crabs, shrimp, prawns, and crayfish. Almost all crustaceans live in water—either seawater or fresh water. Pillbugs are an exception. In other words, those pillbugs, woodlice, or sowbugs you see in the damp soil under rocks are land-dwelling relatives of lobsters, crabs, and crayfish, and they get most or all of their oxygen in the same way as those aquatic animals—through gills.

A gill is a special surface exposed to the environment on one side and to the animal's circulatory system on the other. The gill surface allows oxygen dissolved in water to pass from the environment into the circulatory system. That oxygen must be dissolved in water for the gill to work. Gills must be kept wet. Aquatic crustaceans always have water on their gills anyway. But land crustaceans such as pillbugs must make a special effort to keep their gills wet. So most pillbugs prefer damp places.

Pillbugs have gills on their legs, and they have water tubes on their undersides that replenish the water film on those gills. They get water from their food, but they can also drink liquid water if they encounter it. Pillbugs have gills just like their aquatic relatives the lobsters. Incidentally, we humans also use wet surfaces to transfer oxygen into our circulatory system. But our wet surfaces are protected deep inside our bodies, in our lungs.

Curtis, Helena. *Biology*. 4th ed. New York: Worth Publishers, 1983.
Grzimek's Animal Life Encyclopedia. New York: Van Nostrand Reinhold Company, 1984.

Sleet, Hail, and Snow

Here is a very quick lesson on the differences between sleet, hail, and snow. All three have one thing in common: they are various forms of frozen water. Sleet and hail have several similarities, but snow is very different from them.

Sleet is transparent, solid grains of ice that are smaller than two-tenths of an inch in diameter, and spherical or irregular in shape. Sleet is formed when raindrops or melted snowflakes fall from warm air through a lower layer of below-freezing air, where the raindrops freeze and the melted snowflakes refreeze.

In many respects hail is a lot like sleet, but larger. Hailstones range from the size of a pea to the size of an orange, and there have been some recorded even larger than that. On average, though, they are smaller than one inch in diameter. The way hail forms is a good bit different from the process by which sleet forms. Hailstones form in thunderclouds, and begin as frozen raindrops or snow pellets called "hail embryos." The embryos come in contact with super-cooled water droplets, that is, water which has remained unfrozen at below-freezing temperatures.

As the embryos move through the droplets, the water freezes to them, and a hailstone forms as this freezing water accumulates on its surface. Strong updrafts in the thundercloud keep the developing hailstone moving around the cloud until the stone becomes too heavy and falls to the ground. How large a hailstone becomes depends on how long it remains in the cloud.

Snow is tiny ice crystals that grow from water vapor in cold clouds. For the most part snow crystals form in clouds with below-freezing temperatures. Snow crystals always have six sides, and they come in two forms: platelike crystals, which are the familiar star-shaped snowflakes, and columnar crystals, which resemble six-sided needles of ice.

"Hail," "Sleet," and "Snow." In *New Britannica,* 15th ed. Chicago, 1991.
"Hail," "Sleet," and "Snow." In *The World Book Encyclopedia,* vols. 9 and 17. Chicago, 1994.

Radar

What comes to mind when someone mentions radar? Tracking down enemy missiles, navigating a plane or ship, a speed trap, a character in *M*A*S*H?* Well, radar is involved in all these things. It has been around as long as most of us can remember, and has been invested with almost magical powers, as with the character in *M*A*S*H* who could almost see into the future. But basically radar is straightforward and simple, and considerably more limited than the myths about it.

The word "radar" is an acronym for Radio Detecting and Ranging. It works pretty much like an echo we hear, but instead of sound waves it uses radio waves. You can illustrate for yourself how radar works by using the echo analogy. All you need is a stopwatch and a friend to make a sharp, loud noise, a handclap, for instance. Stand several hundred feet from a reflecting surface, such as a wall. Have your friend clap her hands, and at the same time you start the watch. When you hear the sound echo back, stop the watch. Now multiply the time on the watch by the speed of sound, which is 1,200 feet per second, and divide by two. The result tells you about how far you are from the reflecting surface.

Radar works the same way, except radio waves are transmitted and received instead of sound waves. A radio transmitter beams waves at a distant object, and the time it takes for the waves to be reflected back to the receiver tells how far away the object is—just as with the echo. It all takes place a lot faster, though, since radio waves travel at the speed of light—186,000 miles per second.

So radar can tell you how far you are from something, such as another airplane, but it cannot tell you *where* you are. There are devices which can do that, and one such system is called LORAN, which is the acronym for Long Range Navigation.

LORAN is similar to radar, but instead of using the echo principle, it beams a precisely timed signal from a transmitter to a receiver, and rather than just one transmitter there are three. The transmitters are land-based and spaced some distance apart so that their signals will be received by, say, a plane or a ship at different times. A computer calculates the difference in the time it takes the signals to reach the receiver on the plane or ship, and pinpoints its location.

To see how this works, let's put ourselves on a ship equipped with a special receiver and computer. We're sailing up the East Coast, and want to know where we are. You'll be the navigator. We'll use hypothetical LORAN transmitters in New York City, Trenton, New Jersey, and Boston to guide us. Our receiver gets the signals from the three stations and computes the distance we are from each of them. When we take our readings, we find we are 94 miles from New York City, 127 miles from Trenton, and 133 miles from Boston.

As navigator you take down the distances. Now, using New York, Trenton, and Boston as centers, you draw circles on your map representing those distances. When you have drawn all three circles, there will be one spot where they all intersect at the same point, and that is where you are—in this case, 27 miles off the coast of Long Island, New York. In actual use, the electronic navigation system draws the circles, so to speak, but the principle is the same.

Gutleben, Glenn. "High Tech on the High Seas." *Exploratorium Quarterly*, Spring 1991. *McGraw-Hill Encyclopedia of Science and Technology*. 6th ed. New York, 1982.

Blinking

On average we blink about 14,440 times a day. Since each blink takes about a quarter of a second, about an hour of our waking hours is spent with our eyes partly or completely closed. It has been only in recent years that scientists have really begun to understand blinking. Probably most of us think that the main purpose of blinking is to clean and lubricate our eyes, but it's not. We blink about 15 times a minute, but only one or two of those blinks are needed to keep the surfaces of our eyes rinsed and moistened. We also blink a few times a day because of dust or smoke in our eyes, or being startled. But for the most part blinking is an indication of what's going on in our brains.

Generally speaking, the harder we concentrate, the less we blink. Car drivers blink less in city traffic than on the open road; they probably won't blink at all while passing a truck at high speed.

Further insights into blinking have been gained from studying subjects who were reading. They blinked most often at a punctuation mark, or the end of a page. It was sort of like a signal that the brain was taking a break, a mental punctuation mark, as it were.

The concentration connection wasn't evident just in visual activities. Someone who is anxious tends to blink more than one who is calm. A steady gaze is associated with confidence and self-assuredness; TV anchors are instructed not to blink much, so as to give the impression of being in control.

If blinking is a sort of mental punctuation, this might explain why people solving mental arithmetic blink at different rates. Some do not blink until they have mentally solved the problem, while others blink with each step in the process of solving it. If you want to test this, watch people on the sly so they don't become conscious of blinking, and although you won't be able to see *what* they are thinking, you might get a glimpse into *how* they are thinking.

Ingram, Jay. *The Science of Everyday Life.* New York: Penguin Books, 1989.
Stern, John A. "What's behind Blinking." *The Sciences,* November–December 1988.

Coriolis Effect

If you were to try to fly a plane in a *straight line* from Chicago to Atlanta, you would never get there because you would fly to the right of Atlanta. That doesn't seem logical, but it's true. This sideways drifting is called the Coriolis effect, and it's caused by the rotation of the earth. Although those of us who are earth-bound don't really have to be concerned about it, airplane pilots have to make adjustments.

To understand how the Coriolis drift affects planes, you need to be aware that if you're standing on the equator, the earth's rotation is carrying you eastward at about 1,000 miles per hour. As you move away from the equator, your speed decreases. For example, Boston is traveling at about 700 miles per hour.

Now back to our attempt to fly from Chicago to Atlanta in a straight line. As our plane sits in Chicago, it is spinning with the earth, of course, and when it takes off it continues to do so. However, our destination, Atlanta, being closer to the equator, is moving *faster* to the east. So if we do not correct for the faster eastward movement, we will end up to the right of Atlanta.

(Incidentally, because Atlanta is moving eastward faster than Chicago, if you try to fly northward in a straight line from Atlanta to Chicago you will still end up to the right of your destination.)

This sideways drifting does not occur just with airplanes, it applies to everything that moves on earth. Were it not for the friction of the tires on the road, a car traveling down a highway at 60 miles per hour would be carried off the road to the right at the rate of about 15 feet per mile.

This Moment of Science has dealt only with how the Coriolis effect applies to navigation, but, among other things, it affects weather and ocean currents as well. But as the saying goes, that's another story.

Ingram, Jay. *The Science of Everyday Life.* New York: Penguin Books, 1989.
McDonald, James E. "The Coriolis Effect." *Scientific American*, May 1952.

Illusion in a Coffee Cup

Things are not always as they seem, and this little demonstration will prove it. All you need is a cup of black coffee and an overhead light. A single incandescent bulb works best. Position the coffee cup so that the light is reflected in it. Look into the cup from a distance that allows the reflected light to just about fill the cup.

Now move your head quickly and smoothly toward the cup. The light appears to get smaller and farther away! The change is dramatic. The light seems to shrink to a quarter or a fifth of its original size, and to move away 10 times more than you moved.

This is obviously an illusion, since you know the light hasn't changed size or moved. However, as you move closer to the cup, the cup fills more of your field of vision than the light, so the cup seems larger, and the light smaller.

In trying to interpret the information being sent to it, your brain correlates smaller with farther away, and you perceive that the light has moved. Your brain cannot make an adjustment either; you can do this over and over again and the result will be the same.

Ingram, Jay. *The Science of Everyday Life.* New York: Penguin Books, 1989.
Senders, John. "The Coffee Cup Illusion." *American Journal of Psychology* 79 (1966).

Knuckle Cracking

How many of us have derived impish delight from annoying our mothers by cracking our knuckles? It's not something I did a lot, but I do recall my mother trying to discourage me by telling me it would cause my joints to swell. But then again, she also told me my hair would fall out if I wore my hat indoors.

While the annoyance factor related to knuckle cracking is indisputable, the actual cause of the noise was a puzzle until the early '70s. The most common explanations for the noises were bones snapping against each other, or tendons moving over bony projections in the joints. It kind of gives you goose bumps just thinking about it, doesn't it? Well, this Moment of Science is going to tell you what really makes the sound of cracking knuckles.

The sounds come from tiny explosions. Now, if the thought of your joints exploding doesn't make you feel any better, let me clarify. The sounds are not your joints exploding, but the popping of gas bubbles which are in the lubricating fluid that fills the joint.

In order to crack your knuckles you stretch the joints, causing an increase in the space between the finger bones. This increase in space reduces the pressure on the fluid that lubricates the joints. This reduction in pressure causes tiny gas bubbles to form in the fluid. As the pressure continues to go down, the bubbles burst, making the popping noise you hear.

After the bubbles burst, the gas does not escape the area, but is reabsorbed into the fluid as the joint returns to its original position. It takes about 15 minutes for the fluid to be reabsorbed, and that's why once you crack your knuckles, you can't do it again for a while.

So, even though the evidence indicates that you will not suffer disfigurement from cracking your knuckles, for the sake of mothers past, present, and future, if you feel the need to pop your knuckles, don't do it around your mother—or anyone else's.

Time, August 16, 1971.
Walker, Jearl. *The Flying Circus of Physics with Answers.* New York: John Wiley and Sons, 1977.

You're Not Holding Your Tongue Right

How many times have you noticed someone's tongue peeking out from between the teeth or lips when he or she was concentrating hard? You know, the classic tongue-in-the-corner-of-the-mouth look. Scientists have given this form of expression the unimaginative name of "tongue-showing," and it is a common and powerful form of non-verbal expression.

Tongue-showing is an unconscious act, so it does not include sticking your tongue out at someone. Tongue-showing comes in many forms, such as sticking it slightly between your teeth or lips, curling it inside your mouth and holding it there, etc. Whatever form it takes, it conveys one message: "Leave me alone." The fascinating thing about this is that psychologists have found that neither the tongue-shower nor the tongue-showee is aware that this message is being sent or received.

One interesting experiment which demonstrated that the "leave me alone" message was being sent involved 50 students who were given a test with one page missing from the test booklet. When they discovered the missing page, they approached the instructor to ask him about it. The teacher was ostensibly absorbed in his work, and when the students approached, he did not acknowledge them right away, but did show his tongue to half of them but not the other half.

A hidden observer timed how long it took each student to make a move to get the instructor's attention. The researchers found that the students to whom the tongue was shown took an average of *two and one-half times longer* to interrupt the teacher. Even more intriguing is that not one student had realized that the teacher's tongue had been shown, although each one sensed that the teacher did not want to be interrupted. In contrast, those to whom the tongue was not shown felt no reluctance at interrupting the teacher.

So, next time you're concentrating hard on a task and you get interrupted, maybe you aren't holding your tongue right!

Ingram, Jay. *The Science of Everyday Life.* New York: Penguin Books, 1989.
Smith, John W.; Julia Chase; and Anna K. Leiblich. "Tongue-Showing: A Facial Display of Humans and Other Primate Species." *Semiotica* 11, no. 3 (1974).

Why Are Clouds White?

We've all noticed the color of clouds, which ranges from brilliant white to almost black. Before we consider why clouds have such a range of color, we need to think about the nature of light. When all visible wavelengths of light reach our eyes, we see white. When only certain wavelengths reach us, we see the corresponding colors. When no visible wavelengths strike our eyes, we see black. For most of the day the sun radiates all visible wavelengths of light, and it appears white.

When the sunlight reaches a cloud, some of it is reflected away from the earth. The light which passes through the cloud gets scattered by tiny cloud droplets in all directions more or less equally. This scattering is something like a pinball being bounced off the pins and bumpers; when the light from the sun hits small particles in the atmosphere, it's knocked around. Since the sunlight is scattered fairly evenly by the cloud, allowing all wavelengths to reach us, the cloud looks white.

As a cloud grows larger and taller, more light is reflected from it, and less light is able to penetrate it. When a cloud gets a little over 3,000 feet thick, very little light is able to pass through it, and the base of it looks dark. At the same time the water droplets at the base of the cloud get larger, and as they do they absorb more light than they reflect. Now even less light gets through the cloud, and it appears almost black. These large, light-absorbing droplets often get heavy enough to fall from the cloud as rain. Just from casual observation we know that these dark, ominous clouds often bring rain. Now we know why they look so dark.

Ahrens, C. Donald. *Meteorology Today: An Introduction to the Weather, Climate, and the Environment.* St. Paul: West Publishing Company, 1991.

Oak Trees Outwit Mice

In a mythical Virginia newspaper we might call the *Acorn Times*, the banner headline reads "Oak Trees Outwit Mice." In the Appalachian Mountains of Virginia, it seems that the oak trees synchronize their production of acorns in a way that lets them stay one step ahead of the mice. In years when the white-footed mouse population is low, acorn production is high. Then in years when there are a lot of mice, the oak trees produce very few acorns. It looks as if the trees are pulling a fast one on the mice to keep the rodents from eating all their seeds.

It works like this: When the mouse population is low, the trees produce lots of acorns, many times more than the mice could possibly eat. With the help of this bountiful food supply, there is a mouse population explosion. Then for the next three or four years the trees produce very few acorns, if any. This lack of food causes a marked decline in the mouse population. The drop in the number of mice is followed by an increase in acorn production. And so the cycle goes. This strategy, if that is what it is, results in the survival of more seeds to germinate.

It is not clear how the trees synchronize. It could be an as yet unknown chemical which passes from tree to tree triggering acorn production, or a group response to climate fluctuations. It's more likely that the process evolved by chance: Trees with more acorns in one year than the next produced more offspring over time than those with smaller but consistent crops each year which fell prey to predators. It could just be coincidence, but it sure is to the oak trees' advantage.

"Confounding the Rodents." *Discover*, November 1992.

Let Me Sleep on It

Research has given new meaning to the phrase "Let me sleep on it." Students have sometimes found that after a good night's sleep they are able to remember very well material they studied the night before. Although this often is the case, you can't count on it. Some scientists think that sleep does indeed have an effect on memory, and to try to get a handle on these effects they have done experiments involving different stages of sleep.

Sleep is divided into five stages, including the one called the rapid eye movement, or REM, stage. In the course of a night's sleep, you go through the various stages several times, but it is only during REM sleep that dreaming occurs, and it is this stage which has the most profound effect on memory. You can recognize when someone is in REM sleep, because you can see their eyeballs moving underneath their closed eyelids.

In one study subjects were trained to recognize several patterns of lines a few hours before they went to bed. The experiment found that the volunteers could perform the tasks faster the next morning if they had had a good night's sleep. If they were awakened each time they entered REM sleep, they did no better than the night before. If instead they were awakened during non-REM sleep, they performed just as well as when they slept undisturbed.

The researchers concluded that REM sleep, and perhaps even dreaming itself, helps cement things in memory. The catch is that your REM sleep must not be disturbed, and controlling that is not as simple as hanging a "do not disturb" sign on your door. You can be awakened from REM sleep by dreams as well as a knock at the door.

At any rate, it could well be that if you want to increase your chances of remembering something, study it before you go to bed.

Ezzell, Carol. "For a Good Memory, Dream On." *Science News*, November 14, 1992.

Dimples in Golfballs

You're out on the golf course one pleasant afternoon. Your ball is set on the tee. You lean over the ball. You grip the club just right. Your arms are straight. Your stance is perfect. Whoosh! You swing, and the ball takes off toward the green, sailing in a beautiful arc. Your impeccable technique certainly has a lot to do with the success of that drive, but you got a major assist from the little dimples on the golf ball. Combined with the proper spin, the dimples help keep the ball in the air longer, and here's how.

Your picture-perfect swing puts backspin on the ball. The dimples trap a layer of air next to the ball, and this layer spins with the ball. The air being dragged across the top of the spinning ball moves in the same direction as the air that's rushing past. As the air spinning with the ball comes around the bottom, it is moving in the opposite direction from the air on top, and therefore against the onrushing air. Consequently it's slower than the air on top. Once again we encounter Bernoulli's principle: A slow-moving air stream has higher pressure than a fast-moving air stream. So the higher pressure on the bottom of the ball is going to hold the ball up longer. The effect of the dimples is so significant that a drive of 200 yards hit with a dimpled golf ball would be shortened with a non-dimpled ball to about 100 yards.

Incidentally, with top spin, which is the reverse of backspin, the principle is the same, but the effect is just the opposite—and disastrous for the golfer: with top spin the high pressure is on the top of the ball, so after a short flight the ball takes a nosedive into the ground.

Flatow, Ira. *Rainbows, Curve Balls and Other Wonders of the Natural World Explained.* New York: William Morrow and Company, 1988.

Where's the Plane?

How many times have you heard a jet plane going over, looked up to see it, but not been able to find it even though you were looking toward where the sound was coming from? Sometime during your life you might have been told that the plane wasn't with the sound because it was traveling faster than the speed of sound. It might have been, but it didn't have to be going that fast in order to appear to be separated from the sound.

The reason the jet appeared to be apart from the noise it was making was simply that sound travels a lot slower than light—about a million times slower. For quick review, at sea level sound travels about 750 miles per hour, while light travels at 186,000 miles per *second.*

Commercial jets normally fly at about 30,000 feet. From that height it takes sound about 30 seconds to reach earth. Light gets here in only about four one-hundred-thousandths of a second, or for practical purposes, instantaneously. A plane traveling at 600 miles an hour goes 5 miles in the 30 seconds it takes for the sound to reach you. By the time the sound from the plane reaches your ears, the plane is 5 miles from where that sound originated. So when you look toward the sound, the plane is not there anymore.

Vergara, William C. *Science in Everyday Life*. New York: Harper and Row, 1980.

Tiny Bubbles

In soft drinks, champagne, or other carbonated drinks, we've all seen the little rows of bubbles streaming from the sides of the glass. The sequence of events leading to the streams of bubbles starts with the glass itself. The sides of the glass look smooth, but actually there are tiny nicks and imperfections in the surface. As you pour the drink, a tiny bubble of air is trapped in a nick. Molecules of air attract molecules of carbon dioxide, commonly known as CO_2 (the gas which makes the drink fizzy). The air molecules sort of say "come hold hands with me," and the CO_2 can't resist the invitation.

However, soon so many CO_2 molecules gather in the nick that they form a bubble which is too buoyant to be held there, and it floats to the top. As in life, there is always somebody waiting to take its place, so more bubbles form in the same way, resulting in a continuous stream coming from the nick in the side of the glass. Even the smoothest glasses are covered with tiny nicks, and this explains why there are lots of streams of tiny bubbles.

Flatow, Ira. *Rainbows, Curve Balls and Other Wonders of the Natural World Explained*. New York: William Morrow and Company, 1988.

Night Vision

When we are in a fairly dark room, or outside at night away from lights, we can still see, but we can't see the colors of things very well. Why is that? To find out, let's take a "look" at night vision.

There are two kinds of light-sensitive organs located in the backs of our eyes: rod-shaped and cone-shaped. Both rods and cones are sensitive to light. The difference between them is that the rods allow us to see in very dim light but don't permit detection of color, while the cones let us see color, but they don't work well in dim light.

When it gets dark, the cones lose their ability to respond to light. The rods continue to respond to available light, but since they cannot see color, so to speak, everything appears to be various shades of black and white and gray.

A curious thing is that in dim light you can see more clearly out of the side of your eye, because the light-sensitive rods are more highly concentrated off to the side in the back of your eye. So next time you're out on a clear night, notice how little color you can see, and how you can see objects like dim stars better out of the corner of your eye than from the center.

Flatow, Ira. *Rainbows, Curve Balls and Other Wonders of the Natural World Explained.* New York: William Morrow and Company, 1988.

112 *Alcohol as an Antiseptic*

Alcohol is such a good germ killer because it has the ability to coagulate a germ's protein. The coagulation of the protein is somewhat like the coagulation of your blood on a cut. You can see how if all your blood became hard like that you would not be a very effective living specimen. Alcohol has the same effect on germ cells, and the cells die, too.

Oddly enough, a solution of 70 percent alcohol, that is, rubbing alcohol, is a more effective antiseptic than 100 percent alcohol—another case where more is not better. Pure alcohol doesn't work as well because when it contacts a germ cell it coagulates the cell's outer wall immediately. This forms a protective shell around the cell, which keeps the alcohol from getting through to the protein inside. So even though the bacterium becomes ineffective, it's not dead, and under the right circumstances it can function again.

Rubbing alcohol is not as concentrated as pure alcohol, so even though it coagulates protein just as well, it acts more slowly. This slower action gives the alcohol time to soak into the complete bacterial cell before coagulation takes place. Consequently all of the cell's protein is coagulated, and the bacterium dies.

So when you want to disinfect something, reach for the

rubbing alcohol, because by taking just a little more time to work, it does the job better.

Vergara, William C. *Science in Everyday Life*. New York: Harper and Row, 1980.

What Our Pupils Tell Us

The pupils in our eyes do more than control the amount of light entering them. The size of our pupils can give clues to our interest, emotions, attitudes, and thought processes.

If you've seen a magician tell which card someone picked from a deck, you've probably been amazed. Well, the pupils often open up when the magician picks the right card, so one of the tricks he or she can use is to look at the person's eyes.

The pupils tend to close in response to unpleasantness, and open in response to pleasure. For example, for those who find snakes loathsome, a picture of one will cause their pupils to close. For someone who likes snakes, the same picture will cause the pupils to open.

The same thing happens to people with strong political preferences. A conservative's pupils will close when shown a picture of a liberal. The liberal has the same reaction when shown a picture of a conservative. Their pupils open when looking at a picture of someone of their own persuasion.

When you are concentrating hard on something like a math problem, your pupils will remain open until the problem is solved.

What these and other studies have shown is that our pupils are very sensitive to our emotions and thought processes, often revealing things which we choose not to verbalize.

Vergara, William C. *Science in Everyday Life*. New York: Harper and Row, 1980.

Lightning Fertilizer

We don't usually think of lightning as having anything to do with nutrition, but it does. Our bodies need protein, and proteins contain nitrogen. The air we breathe has plenty of nitrogen to satisfy our needs, but that nitrogen is not available to us directly from the air. The only way we can get nitrogen is from the plants we eat, or from the animals we eat that eat the plants.

But there's a problem. A nitrogen molecule in the air consists of two atoms which are held together very tightly. In order for us to absorb nitrogen, the two atoms must be separated. But the two atoms are held together so tightly that our body chemistry does not have enough energy to process them.

This is where lightning comes in. Obviously we don't have to be struck by lightning in order to satisfy our need for nitrogen! However, in a thunderstorm there is enough electrical energy in lightning to separate the nitrogen atoms in the air. Once the atoms are separated, they can fall to earth with rainwater and combine with minerals in the soil to form nitrates, a type of fertilizer.

The nitrogen-containing nitrates in the soil are absorbed by the plants, and when we eat the plants or the animals that eat the plants, we get the nitrogen in a form which our bodies can use. So, in addition to providing a spectacular light show, and scaring us to death, lightning also helps fertilize the soil.

Vergara, William C. *Science in Everyday Life.* New York: Harper and Row, 1980.

Seeing Stars

At some point in your life you might have bumped your head so hard that you saw stars. Little lights didn't actually flash at the moment of impact, but your brain was tricked into thinking they did.

To understand how the brain gets fooled, it helps to know how we normally see. In the back of the eye there are about a hundred million nerve cells. When light strikes these nerves, they react by sending impulses to the brain. So vision really takes place in the brain as the result of nerve impulses from the eye interacting with signals produced in the brain.

A sharp blow to the head can cause the nerves in the eye to fire off messages to the brain, just as they do in response to light. Since the signals are coming from the eye, the brain interprets them as quick flashes of light, and that's what we see. So the stars we see when bumping our heads are an illusion caused by our brains' being confused by the signals coming from the nerves in our eyes.

Vergara, William C. *Science in Everyday Life.* New York: Harper and Row, 1980.

Where Do All Those Calories Go?

While part of the answer to the title question is "to our hips, thighs, and midsections," that's not where most of them go. In an average day we burn more than 2,000 calories. Surprisingly, most of them are *not* used up in physical activities such as walking or exercising. Most of the calories we take in are used to maintain our body temperature at a fairly constant level. So we and other mammals need frequent meals in order to keep our furnaces stoked, so to speak.

Our temperature varies only about one degree over the course of a day, so why is it so important to keep our temperature so consistent? It's partly because of the effect that temperature has on chemical reactions in our bodies. Although the rate of all chemical reactions speeds up as the temperature rises, the reaction rates for different chemicals do not always increase equally. These unequal increases in reaction rates could result in harmful chemicals building up in our bodies. Maintaining a consistent temperature has allowed the evolution of precise coordinations of different chemical reactions in our bodies.

You're probably aware that internal body temperatures that are too cool or too warm can be harmful or even fatal. Our internal systems have evolved to function in a very narrow temperature range, and if that range is exceeded there can be malfunctions. For example, if you get severely overheated you can suffer from heat stroke, a potentially fatal condition if you are not cooled off quickly.

So the calories from the "three square meals" we are advised to have each day provide a regular source of fuel for keeping our temperature constant.

Kalat, James W. *Biological Psychology.* 4th ed. Belmont, Calif.: Wadsworth Publishing Company, 1992.

Diamonds

You walk into a fancy party. What is the oldest thing there? No, not your in-laws, or even your grandparents. It will be a diamond someone is wearing. More on the age of diamonds later, but first some more amazing mysteries and facts about diamonds.

It's not completely clear how natural diamonds are formed, and although it is known that they are made from carbon, scientists are not sure how carbon gets to the extreme depths at which diamonds are made.

Diamonds are formed very deep in the earth. The optimum depth for their formation is about 120 miles beneath the surface, in the molten mantle. The temperature and pressure necessary for diamonds to crystallize are mind-boggling: the temperature must be over 2,000 degrees Fahrenheit, and the pressure must be at least 690,000 pounds per square inch. To put that pressure in perspective, a 150-pound person exerts only about 3 pounds per square inch. We would never know about diamonds were it not for volcanic activity, for that's what delivers diamonds from deep inside the earth to where we can get to them.

Theoretically, diamonds can remain diamonds only at high temperature and pressure. In theory, at atmospheric pressure and lower temperature, chemical changes are liable to take place that change diamond into graphite, similar to the stuff pencil leads are made of. However, it has been calculated that even if this change were to take place, it would take more than 10 billion years to happen.

Now back to the diamond's age. Scientists think that diamonds may have been forming throughout earth's history. Many have been found that are 3.3 billion years old. And the young ones are a mere 1 billion years old.

Kirkley, M. B.; J. J. Gurney; and A. A. Levinson. "Age, Origin and Emplacement of Diamonds: A Review of Scientific Advances in the Last Decade." *CIM Bulletin*, January 1992.

Legrand, Jacques. *Diamonds Myth, Magic, and Reality.* New York: Crown Publishers, 1980.

Wilson, A. N. *Diamonds from Birth to Eternity.* Santa Monica, Calif.: Gemological Institute of America, 1982.

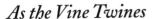

As the Vine Twines

A vine twined around a tree limb or a trellis is a familiar sight, but did you ever wonder why and how it gets that way? The purpose of the twining is to anchor the vine. In some cases the vines themselves do the twining, while in others the vines have small appendages called tendrils that coil around twigs or trellises.

How does the vine know to twine, and in what direction? The fact is, the vine doesn't know anything; its coiling is a blind reaction, so to speak, caused by chemical changes inside it. When the growing tip of the vine touches an obstacle, such as a twig, this stimulates the cells of the vine on the side opposite the twig to grow faster than those next to the twig. As the side of the vine opposite the twig gets longer, and the side touching the twig grows very little, the vine is pushed around the twig.

Most vines are not particular about which way they twine. If the outside of an already twining vine's tip touches an obstacle, the vine will change direction.

For This You Need a Doctor?

Chicken soup is a time-honored treatment for fevers and coughs, recommended since ancient times by physicians and grandmothers. At least one modern scientist daring to investigate the validity of this ancient advice has found it to be correct.

Stephen Rennard at the University of Nebraska Medical Center wanted to see if there was a scientific basis for the chicken soup remedy. He knew that when the body is invaded by a virus or harmful bacteria, it sends certain white blood cells to release enzymes to help fight the infection. Unfortunately the enzymes that fight the infection also irritate our tissue, and give us the sore throats that so often come with the flu.

Rennard devised an experiment in which he could observe the movement of white blood cells toward bacteria, similar to the movement that takes place in the human body. He hoped to test what effect, if any, chicken soup had on this movement.

Using a recipe from his wife's grandmother, he added the various ingredients of the soup in stages to a container that held white blood cells and bacteria. Plain water had no effect on the movement of the white cells toward the bacteria, but when he added the vegetables, the cells moved distinctly slower. Unfortunately, the vegetable broth also killed some of the body's white cells.

However, when he added chicken to the vegetable soup recipe, no harm was done to the cells. Evidently the chicken counteracted the toxic effects of the vegetables. The cells' movement was still slowed, so Rennard assumed that in a human body the slowing down may reduce the number of enzymes at the site of infection, and probably reduce the inflammation and some of the discomfort. Interestingly, the slowed movement did not seem to affect the cells' infection-fighting abilities. Also, this bit of research showed that although a remedy might be traditional, that doesn't mean there isn't a scientific basis for it.

Magic Numbers

Here's a number puzzle that's bound to make your friends think you are either a mind reader or a magician. Ask your friends to write down any two different numbers from 1 through 9 without letting you see them. Now tell them to reverse the two numbers. They now have two, two-digit numbers. Tell them to subtract the smaller number from the larger one.

Now they should take the number that resulted from the subtraction, reverse the digits, and add that number to the one they got when they subtracted. When they are finished you can tell them the answer is 99. They can do this with any combination of numbers with two different digits from 1 through 9, and you will always be able to tell them the answer, because it will always be 99.

Any time this five-step process is followed with two-digit numbers from 1 to 9, the result will be 99. For example, suppose you choose 8 and 6 as your two digits. Eighty-six reversed is 68; 86 minus 68 is 18; 18 reversed is 81; and 81 plus 18 is 99.

Incidentally, a practical application of this phenomenon is that if your checkbook doesn't balance, and the difference between your balance and the bank's is evenly divisible by 9, there are probably two transposed numbers somewhere.

Note: This same procedure can be followed with three-digit numbers; the answer then is 1,089.

Where Are the Poles?

Here's what seems like an easy test of your knowledge of geography. Stand facing west, and simultaneously point to the north and south poles with your right and left arms. Now, are your arms pointing straight out from your body in opposite directions? If they are, they are not pointing toward the poles, but to outer space. It's the roundness of earth that causes your pointing straight out from your body to the poles to be wrong. In order to point at the poles, your arms must point not only northward and southward, but downward.

To see this clearly, draw a circle on a piece of paper; the circle represents the earth. Make marks at the top and bottom of the circle to indicate the north and south poles. Draw a stick-figure person without arms on the circumference of the circle anywhere but at the poles. Now draw the arms pointing straight out from the body to the right and left, and you can easily see that they are not pointing at the poles, and that to point to the poles they need to be angled downward.

If you draw the arm lines on your stick figure so that they point to the poles, you'll see that a 90-degree angle is formed where the arm lines meet the body. This is the way it is in real life, too, and there's a principle of geometry that explains how this works. Applied to this situation, it basically says that if you could point directly at the north and south poles, and draw imaginary lines from them that come up your extended arms and meet at about your neck, the lines would form a 90-degree angle.

As a matter of fact, no matter where on earth you are, your arms will form a right angle when you point to the poles—except for two places. Do you know where those two places are?

Evans, Barry. *Everyday Wonders: Encounters with the Astonishing World around Us.* Chicago: Contemporary Books, 1993.

Dark Water Spots

When we splash water on our clothes or get caught in the rain, the places where the water hits appear darker than the rest of the cloth. Does the cloth really change color when it gets wet? No, it doesn't, but something does happen to the white light striking the cloth that makes it seem as if there is a color change.

White light, such as sunlight, is made up of a mixture of light of all colors. So when it lands on a pair of blue jeans, for instance, the jeans look blue because when white light strikes the blue threads, mainly blue light is reflected back to our eyes, while most of the other colors are absorbed by the cloth.

When a spot on the jeans gets wet, that area is coated with a film of water, and water fills spaces around the threads. Light striking the wet spot is bent down among the fibers. There it may reflect off the surface of the water, and bounce off the threads several times before it returns to our eyes.

With each bounce more light is absorbed by the cloth, and more non-blue light is absorbed than blue. Eventually some of the light is reflected back to our eyes, and since more of the non-blue light has been absorbed by the wet cloth, what we see looks bluer than the dry material.

The dry cloth looks lighter because there's no water to bend the light down between the fibers where more of it can be absorbed. So even though there is still more blue light reflected, more of all the colors is reflected to our eyes as well. Since all colors combined make white, it's sort of like mixing white with blue to make a lighter blue.

Spotting Satellites

This is a fun activity that you can do during the first two hours of darkness from May through August.

There are thousands of satellites and other pieces of large space junk orbiting the earth. A few hundred of them are visible with the naked eye, and a few thousand with binoculars. You can tell a lot about a satellite from ground observation, but first you need to determine if what you're seeing is a satellite. Any dot of light moving across the sky that's not a plane is almost certainly a satellite. Look for points of light that move from west to east or from north to south or vice versa, because satellites always move in those directions.

When you spot a satellite, there are several things you can tell about it, mainly by judging its speed. If it takes about three minutes for the satellite to cross the sky, it is probably a military spy satellite in a low orbit—about 120 to 180 miles high. If it is moving faster than that, it is probably no longer operational, and is close to being pulled back into earth's atmosphere, where it will burn up.

A slower-moving satellite might be a weather satellite in a higher orbit, especially if it moves from north to south, or vice versa, in what is called a polar orbit. And again, a faster-moving one in a polar orbit might be a spy satellite. Polar orbits are preferred for weather and military observations because the satellite can scan the whole earth in one day.

Incidentally, the satellites that backyard dishes are pointed at can't be seen from earth because they are 22,300 miles out in space.

Ahrens, C. Donald. *Meteorology Today: An Introduction to the Weather, Climate, and the Environment.* St. Paul: West Publishing Company, 1991.

Berman, Bob. "Satellite Season." *Discover,* May 1991.

Two-Point Threshold

Here's a fun experiment that tells you something about your nervous system. You can do it on yourself, but it works best if you do it on someone else, and vice versa. All you need is two pointed objects, such as pencils.

Have your friend close her eyes, and tell her you're going to touch the inside of her forearm with the two pencil points simultaneously about eight inches apart. Then you're going to lift the points and bring them down again and again, each time moving the points closer together. She's to tell you when you're touching her with only one pencil. It's surprising that when she *thinks* she's being touched by only one point, *both* are still touching her arm, and they are one to two inches apart!

Now do the same thing on the pad at the end of her index finger, but start with the points about an inch apart. This time when she thinks she feels only one point, the two will be about an eighth of an inch apart.

Why does she feel only one point when she's being touched by two? And why are the points as much as two inches apart on the forearm when one point is felt, but only an eighth of an inch apart on the fingertip? It's because some areas of the body, such as the fingertips, have more nerves going to them. And the areas of the brain which receive information from these sensitive areas of the body have a greater density of nerves. So since more nerves are present to detect sensations, these areas are more discriminating.

This is an interesting experiment, but there is a useful side to the information. This psychological phenomenon is known as the two-point threshold, and it can be used to test for nerve damage. Also, Braille dots were designed so that they are farther apart than the two-point threshold.

Geldard, Frank A. *Fundamentals of Psychology.* New York: John Wiley and Sons, 1962.
Schiffman, Harvey R. *Sensation and Perception: An Integrated Approach.* 2nd ed. New York: John Wiley and Sons, 1982.

Some Like It Hot

Most of us have jumped into a swimming pool and felt the shock of the cold water, only to have it feel just fine after a minute or so. Or we've stepped into a nice warm shower and after a minute reached over and turned up the heat a little because it was not feeling warm enough. Those who like hot showers might turn up the heat a couple of times before it suits them. This ability of the body to adjust to temperatures is called thermal adaptation. When you adapt to a temperature, it means it doesn't feel cold or hot, but neutral.

A simple experiment can clearly demonstrate thermal adaptation. Get three bowls large enough to put your hands in. Put cold water in one; you might add a few ice cubes to make the water cold. In the second put water that is hot, but not too hot to put your hands in. And in the third bowl put water that is warm, about 90 degrees.

Put one hand in the bowl with the cold water and the other in the hot water for about a minute. Now put both hands in the bowl of warm water. The warm water will feel very cool to the hand that originally was in the hot water, and warm to the hand that was in the cold water.

Fortunately thermal adaptation has its limits, because otherwise we might get burned or frozen if at some point the warning sensation of extreme heat or cold didn't override adaptation. Scientists do not completely understand the process of thermal adaptation, but they've known about it for a long time. The first report of the experiment with the three bowls of water was given by the philosopher John Locke in 1690.

Geldard, Frank A. *Fundamentals of Psychology.* New York: John Wiley and Sons, 1962.
Schiffman, Harvey R. *Sensation and Perception: An Integrated Approach.* 2nd ed. New York: John Wiley and Sons, 1982.

When You Eat an Egg, Are You Eating a Baby Chick?

Don Hirose of Honolulu, Hawaii, and a friend had a friendly argument. Don's friend was sure that when you eat an egg you're eating a baby chick, but Don didn't think so. They made a bet and came to *A Moment of Science* to have it settled.

Before we tell you who has to pay up, let's examine the two situations we're dealing with. The first, and less common, scenario involves eggs that are bought from small farmers—so-called free-range eggs.

These often are fertilized eggs, and can develop into baby chicks. But usually the development process is halted before this happens. When the egg first forms, it's only one cell, and is fertilized as it moves down the oviduct to be laid. During the first 24 hours after fertilization, the single egg cell divides a few times, forming a little mass of cells. At this point it's technically an embryo (though it doesn't look like a baby chick), but the cells still have not separated into the ones that make eyes, feet, feathers, etc.

After the egg is laid, the embryo stays in a kind of suspended animation until the hen sits on it to incubate it. If the egg is not incubated within a few weeks, the embryo will die. This is how things work with a fertilized egg, the kind one might get directly from a small farmer.

The second scenario involves the eggs we buy at the local supermarket, which come from what are called "egg factories." The egg factories rely on a peculiarity of hens: they lay eggs whether or not they're fertilized. The egg factories do not have roosters in residence to fertilize the eggs, and these eggs will not develop into baby chicks.

So the eggs that most people eat do not have embryos, and even the egg cells in the "free-range" eggs probably have not developed enough to be at the stage where one would be eating a baby chick. Therefore, Mr. Hirose wins the bet, and maybe he'll want to buy an omelet with his winnings.

Alcoholism

Young men often brag about how they can "hold their liquor." In fact, many young men do have a high tolerance for alcohol. Unfortunately, this might be a bad omen.

Some young men can drink as many as three to five alcoholic drinks with few, if any, intoxicating effects. However, research by Dr. Marc Schuckit of the VA Hospital and University of California, San Diego Medical School indicates that this trait increases the chances that a man will become an alcoholic by age 30. There is an even greater chance if the man's father was an alcoholic, but the chances are also enhanced for men with non-alcoholic fathers, although to a lesser degree.

The primary factor, though, is a low sensitivity to moderate amounts of alcohol. This means that the man has to drink more to feel intoxicated, and his alcohol tolerance might lack the warning signs that tell him to stop drinking. He might feel little or no effect from three to five drinks, but then become drunk rapidly after reaching a threshold level of imbibing.

Even if a man does not set out to get drunk, the fact that he needs to drink more and more to feel even moderate effects can create a problem. The increasingly greater intake of alcohol can lead to physical and psychological dependence on the drug.

The research does not indicate that low sensitivity to alcohol is a specific cause of alcoholism. However, it does give strong evidence that this trait increases the chances of developing alcohol dependence. It can be inferred from these studies that young men with a high tolerance for alcohol should exercise caution with their drinking habits.

Peterson, Ivars. "Alcoholism Exposes Its 'Insensitive' Side." *Science News*, February 19, 1994.

Schuckit, Marc. "Low Level of Response to Alcohol as a Predictor of Future Alcoholism." *American Journal of Psychiatry*, February 1994.

Soap vs. Detergent

Soap has been used for cleaning for thousands of years, but it was not until modern chemists began to understand its molecular structure that anyone knew *how* soap worked its magic. The long soap molecule has one end that is attracted to fats and oils; the other end is attracted to water. When soap is added to the wash water, one end of its molecule attaches to the oily dirt and pulls it away from the fabric or your skin. The other end stays attached to the water, and when the water is washed down the drain, the dirt attached to the other end of the molecule follows.

The problem with soap is that it doesn't work well in hard water. Hard water contains a lot of calcium, and before soap begins to clean you or your clothes, it separates the calcium from the water. This is what makes the scum of the bathtub ring. After the soap has removed all the calcium in the water, it starts to clean. That's why it takes more soap to clean in hard water—the first soap gets rid of the calcium, then more is needed to get rid of the oily dirt.

After World War II, washing machines became very popular, resulting in large demands for soap. However, the public wasn't satisfied with the grungy film left on clothes. When chemists began working on a cleaner that wouldn't leave a film, they knew they needed to keep the basic structure of soap, that is, a molecule with one end that was attracted to oil and the other attracted to water. To eliminate the film, they developed a substance whose water-attractive end would not have an affinity for calcium. These detergents did not separate the calcium which formed the ring, but left it in to be washed away with the dirt. And these are the detergents we use today.

Collier's Encyclopedia. New York: Macmillan Educational Company, 1991.
Van Nostrand's Scientific Encyclopedia. 7th ed. Edited by Douglas M. Considine. New York: Van Nostrand Reinhold, 1989.

Once upon a Time There Was Air Pollution

126 If this were a fairy tale, it might begin, "Once upon a time there was air pollution"—not because there is no longer air pollution, but because we usually think of industrial air pollution as being a fairly recent problem, beginning perhaps a couple of hundred years ago with the Industrial Revolution. However, some recent research indicates that widespread industrial air pollution goes at least as far back as the ancient Greeks and Romans.

Environmental scientist Ingemar Renberg of the University of Umeå in Sweden sampled sediments from beneath 19 Swedish lakes. He discovered that naturally occurring lead concentrations remained fairly constant for many centuries after the lakes were formed by receding glaciers about 10,000 years ago.

Then about 2,600 years ago, the lead levels in the lakes began to rise. This was when the Greeks began minting silver coins. The silver for the coins was extracted from ore that also contained lead, and the smelting process released some lead into the air.

The ancient contamination recorded in the lake sediments increased again about 2,000 years ago, when the Romans were smelting lead throughout south and central Europe, releasing even more lead into the air. As the Roman Empire declined, so did the lead levels in the Swedish lakes.

Then about A.D. 1000 the lead levels in the lakes began to rise again. At this time the Germans had begun to mine silver and lead. By the beginning of the nineteenth century, the lead measured in the lakes was as much as three and a half times what it had been during Roman times.

Apparently some of the lead that escaped into the air during these ancient times reached Sweden, since the levels of lead in the lakes correspond with the lead-processing activities in south and central Europe. While industrial air pollution is certainly not good, Renberg's research has shown that it's far from new.

"Dumping on the Swedes." *Discover,* July 1994.

Why 5,280 Feet?

Our word "mile" comes from the Latin "mille," which referred to the Roman mile. The Roman mile had military origins, since it was the equivalent of 1,000 double paces of marching soldiers. The soldiers' double paces were about 5 feet, so the Roman mile was about 5,000 feet.

Since we got our measurement system of inches, feet, yards, and miles from the British, what does the Roman mile have to do with our mile? Well, Britain was part of the Roman Empire from the first to the fifth centuries A.D., so when the British began to standardize their measuring system, there was a Roman influence.

Even before the British started keeping written records of landholdings, the farmers laid out their fields in plowed furrows that were consistently the equivalent of a modern 660 feet long. This distance became a standard part of their measurements. Over time, by slurring the words, this "furrow-long" distance became "furlong," a unit that is now used almost exclusively in horse racing.

The British eventually used the Roman mile as a model in their measurement system, but they didn't want to give up their furlong. The Roman mile was about 7.5 furlongs, and when the British adopted it, they lengthened the Roman mile to 8 furlongs, which equals 5,280 feet.

Feldman, David. *Imponderables: The Solution to the Mysteries of Everyday Life.* New York: Quill, 1987.

Are We There Yet?

We're driving to a place we haven't been before. It might be just across town, but it seems like it takes a long time to get there. Then, on the return trip home, even though we travel the same route, it doesn't seem to take nearly as long. Scientists have studied this common phenomenon, and have concluded that our perception of how time passes is sometimes based on the amount of information we're processing. The more information we're getting, the more slowly time passes.

Let's apply this theory to our trip. On the way to the strange place, we're bombarded with unfamiliar sights, and perhaps even sounds and smells. We're not sure where we're going, we're not familiar with the traffic patterns, we're looking carefully for road signs or landmarks, a passenger is reading directions to us, and, of course, the kids are whining, "Are we there yet?" We're constantly having to process and evaluate all kinds of information.

On the return trip home, we're at least somewhat familiar with the territory, so much of the information is not new to us. We now process the information more efficiently. We're able to ignore a lot of it because we remember that it does not require action on our part.

Our mental processes might be compared to one of those little flip-page picture books where you let the pages slip from your thumb and the images seem to move. As you smoothly flip the pages, you get a glimpse of each page, and the images move fairly slowly. If you let the pages slip by in bunches so that you don't see each one, the cartoon action is much faster. In a sense time speeds up. It's much the same with our brains: On the way to the new place we see things in great detail, and time seems to move slowly. On the return trip we're not paying as much attention to detail, so time passes faster.

Friedman, William. *About Time: Inventing the Fourth Dimension.* Cambridge, Mass.: The MIT Press, 1990.

Ornstein, Robert E. *On the Experience of Time.* Baltimore: Penguin Books, 1969.

The Versatile Fruit

If you want to make a jelled salad with pineapple, you'd better not use fresh pineapple, or you'll end up with a soupy mess. On the other hand, only fresh pineapple will work as a meat tenderizer. Five hundred years ago, Christopher Columbus found Indians in the Caribbean using pineapple juice to soften their skin, clean their wounds, remove body hair, and cure upset stomach. The secret of the pineapple is an enzyme called bromelain, which is similar to the enzymes that our own digestive system uses to break down protein.

When you marinate meat in fresh pineapple juice, the bromelain begins breaking down the proteins so that by the time the meat gets to your mouth, the digestion is already begun. Pineapple juice works as a skin conditioner because the bromelain breaks down dead and damaged outer layers of skin, exposing the softer skin underneath. Putting pineapple juice on an open wound might be painful, but it can also break down damaged tissue and kill bacteria.

But one place fresh pineapple does not work is in jelled salads, because gelatin is a form of protein, and so is broken down by bromelain, leaving you with a very sloppy salad. But don't despair; since bromelain is broken down by heat, you can still make a gelatin mold with cooked or canned pineapple—only you won't be able to use the cooked pineapple to tenderize meat, soften your skin, or disinfect wounds.

Bickerstaff, Gordon. "Hidden Powers of Pineapple." *New Scientist,* June 2, 1988.
McGee, Harold. *On Food and Cooking: The Science and Lore of the Kitchen.* New York: Scribner, 1984.

Flipping the Switch for Digestion

The human body is capable of digesting a very wide range of foods, including sugars, fats, and proteins. But the trick to such a varied diet is in the body's ability to stop digesting when the food's all gone so that it doesn't start in on itself.

One of the ways the stomach avoids digesting itself involves the body's careful handling of the strong chemical called protease. Protease is a group of enzymes that break down protein. But since the body itself is made of protein, it's important that those enzymes don't go to work on our own bodies.

The body produces protease in the pancreas, but the pancreas doesn't produce protease in a working condition. Instead, the protease produced in the pancreas has to be activated by another enzyme found in the intestine. Only after it is activated by the other enzyme can the protease go to work breaking down protein. The second, activating enzyme in turn does its job only when food enters the stomach.

At night when there's no food in your stomach, the protease is deactivated so it stops working. Unfortunately, disease, alcohol, and some drugs can override the enzyme that is supposed to be controlling the protease. When that happens, the protease begins to digest the stomach wall and ulcers develop.

So, in a healthy person, the body builds its digestive enzymes with what amounts to an on/off switch and then builds a second enzyme especially designed to operate the switch. The digestive system also protects itself by being one of the fast-growing tissues in the body, constantly discarding old cells and reproducing new ones. So some of it does get digested, but there's always more to take its place.

Bickerstaff, Gordon. "Hidden Powers of Pineapple." *New Scientist*, June 2, 1988.

Creaking Snow

"Probably all, or nearly all, who have experienced a cold winter, are familiar with the cheery cry of the snow as it is pressed against a hard surface by the steel tire of a wagon, for instance, or even onto a pavement by the heels of one's boots." Those words were written many decades ago by the physicist W. J. Humphreys in a book called *Physics of the Air.* Humphreys went on to suggest that creaking of snow is connected with very cold temperatures.

Humphreys said that when the temperature is just below freezing and snow is easily packed into snowballs, footsteps and rolling wheels won't create much sound. His reasoning is based on the fact that applying pressure to ice lowers its melting temperature. If ice is so warm that it's about to melt anyway—say, at a temperature of 30 or 31 degrees Fahrenheit—then a little pressure will be all that is necessary to melt it. Snowflakes are small ice crystals. Snowflakes near their melting temperature can, through hand pressure, be made to fuse into a snowball by a process of melting and refreezing. The snow yields to pressure "gently and progressively" and doesn't make sound.

On the other hand, if the snow is very cold, far below its melting temperature, even the pressure of a boot heel or a wagon wheel won't melt it. The snow is powdery and won't form snowballs. Instead, pressure just makes the ice crystals crush and slip over each other as dry particles. That "abrupt and jerky" motion of the dry ice crystals causes vibration and sound, according to Humphreys's book.

So W. J. Humphreys claimed that snow creaks when the weather is very cold, but not when the weather is only moderately cold. Does that claim jibe with your experience?

Humphreys, W. J. *Physics of the Air*. Reprint of 1940 edition. New York: Dover Publications, 1964.

Now You See It, Now You Don't

In the mid-1990s a new kind of art, called a stereogram, was popular, especially on cards, calendars, and posters. At first these pictures looked like abstract patterns, with wavy lines and bright colors jumbled together at random. Once you relaxed your eyes, however, and stared at the picture in a certain way, you were delighted to see a vivid three-dimensional image suddenly appear. This might have seemed like magic at first, but it had more to do with the way your brain processes visual information.

We see in three dimensions because we have two eyes. Although both of our eyes point in the same direction, they give us slightly different views of the world. You can test this by holding a finger, upright, about a foot in front of your face. Close first one eye, then the other. You will notice that the finger seems to change position as you look at it through different eyes. Your brain takes these slightly different pictures and puts them together, forming a single three-dimensional image. This ability to form a three-dimensional image from two slightly different pictures is the key to how these illusions work.

If you can find one, study it carefully. As you scan from left to right, you will notice that the pattern is quite repetitive. Many features repeat, changing slightly as they recur across the picture. Now relax your focus and try to stare through the picture as though it were a window. The picture will get blurry, then each eye will eventually focus on neighboring features in the pattern. Your brain, seeing slightly different pictures from each eye, will put the images together into a coherent three-dimensional picture. That's when the magic moment occurs, and the colorful, abstract blur resolves into a clear picture. It takes practice to relax your eyes this way, but keep trying. The delightful sensation as your brain switches from two to three dimensions is well worth the wait.

Grimes, William. "Thing; Sleight of Eye." *New York Times,* March 6, 1994.
Rheingold, Harold. *Stereogram.* San Francisco: Cadence, 1994.
Tyler, Christopher, Smith-Kettlewell Eye Research Institute of San Francisco (Dr. Tyler invented stereograms).

Icy Fingers of Frost

After your next hot shower, you might find that your bathroom mirror is covered with a uniform white blur. This is because water vapor from your shower condenses evenly on the surface of the mirror. When water vapor gathers as frost, however, a very different thing happens. Look at the frost on a car window the next cold winter morning, and you could find it organized into intricate and beautiful patterns. Why does water condense evenly on a fogged-up bathroom mirror, but in complex patterns as frost?

It's because there's a difference between the ways water molecules organize themselves as liquid and as ice. The fog that gathers on your bathroom mirror after a shower is composed of minute droplets of liquid water. The molecules in liquid water are free to jostle against each other and shift position. Because of this, liquid water molecules are never very organized. Like grains of sand scattered haphazardly on a beach, this microscopic jumble appears uniform when seen from our perspective.

The molecules in frozen water behave very differently. When water freezes, the molecules are no longer so haphazard. They cling to each other in organized hexagonal patterns.

Frost forms when water vapor condenses out of the air onto surfaces that are below freezing. The first water molecules that stick to the glass freeze in this hexagonal pattern. As more water vapor freezes onto the glass, it attaches itself to the ice that's already there, but only as another hexagonal pattern. Even as it bends and twists along the surface of the glass, frost is always built from these basic six-sided building blocks. When the tempera- ture is right, this rigid organization at the microscopic level leads to the beautiful frost patterns we see. If ice molecules weren't arranged so precisely, frost would always be as uniform as fog on the bathroom mirror.

"Frost" and "Ice." In *New Britannica*, 15th ed. Chicago, 1991.

Healing Elbows and Eyeballs

A scratch to the surface of your eye heals quite rapidly, often in a matter of minutes. But a surface scratch on your elbow takes days rather than minutes to heal. The reason for these different rates of healing is directly linked to the kinds of cells that exist on the outermost layers of your skin and your eyeball. When your elbow or your eyeball is scratched, a layer of cells is actually scraped away from the surface. The healing of a scratch happens when new cells take the place of the ones lost to the minor injury.

When the outer layer of cells on an eyeball is scratched, cellular replacement happens quickly because the surface of an eyeball consists of living cells. These living cells have the ability to rearrange themselves and migrate to where the cells are missing.

The outer layers of skin, however, consist of non-living cells, sometimes layered three or four cells deep. These dead cells create a surface that protects the underlying, more tender, living cells. Because they are dead, the cells that make up the outer layer of your skin can't move like the living cells of an eyeball. In order for new skin cells to replace old ones—in other words, in order for healing to occur—live cells, several layers below the surface, must work their way up to the outermost layer of the skin. When compared to the quick healing that the surface cells of an eyeball can accomplish, the healing of an elbow, which must begin several cell layers below the surface, can take a long time.

Quicksand

It's been a staple of adventure movies and fiction for generations. Our hero is traipsing through a swamp, fighting off all manner of alligators, giant snakes, and whatnot, when—suddenly!—he discovers he's stepped into a pit of quicksand. Can he get free, or will he be drawn to certain death, leaving only his hat, bobbing on the surface, to advertise his fate? If this were real life, not the movies, and if he didn't panic, our hero would have no trouble freeing himself from this peril.

Aside from Hollywood studios, quicksand is usually found near the mouths of large rivers or along beaches and streams where pools of water can become partly filled by sand. It's often formed by a flow of groundwater that passes upward through sand.

Quicksand is a mixture of two things: sand and water. The reason you don't sink into ordinary sand is that the sand grains rest against each other in an interlocking pattern. This mutual contact helps them distribute your weight and hold you up. However, if there's enough water between the grains of sand to keep them out of contact, the sand can no longer bear your weight. The sand and water mixture functions like a liquid, and you start to sink.

Unlike in the movies, quicksand will not suck you under. The mixture of sand and water is denser than water alone, so it's actually more buoyant. You float better in quicksand than in water. The only reason some people have drowned in quicksand is that they've panicked.

Don't ever play in quicksand, but in the unlikely event you're accidentally caught, stay calm, float on your back, and call for help. If no one's around, gently squirm on your back to firmer ground.

Cazeau, Charles J. *Science Trivia from Anteaters to Zeppelins.* New York: Berkley Books, 1986.

Marvels and Mysteries of the World around Us. Pleasantville, N.Y.: Reader's Digest Press, 1972.

"Quicksand." In *New Britannica,* 15th ed. Chicago, 1991.

Why Popcorn Pops

Popcorn, like all grains, contains water. About 13.5 to 14 percent of each kernel is made up of water. So when a popcorn kernel is heated above the boiling point of 212 degrees Fahrenheit, this water turns to steam. The steam creates pressure within the kernel, causing the kernel to explode and turn itself inside out. But if the water inside a piece of popcorn is what makes it pop, why don't other grains pop as well? Wheat and rice contain water, so why don't we sit down to watch a movie with a bucket of popped rice or popped wheat?

The answer lies in the differences between the outer coverings, called hulls, of popcorn and other grains. Unlike rice and wheat, and unlike even regular corn, popcorn has a non-porous hull that traps steam. With the porous hulls of other grains, steam easily passes through, so no significant pressure is produced. These grains may parch, but they will not pop.

But even popcorn, with its special hull, doesn't always pop. Popcorn must have two important properties to pop well. First, the amount of moisture in the kernel must be very close to 13.5 percent. Too little moisture, and not enough steam will build up to pop the kernel. Too much moisture, and the kernels pop into dense spheres, rather than the light, fluffy stuff popcorn fanciers love.

Second, the kernels must not be cracked or damaged in any way. Even a small crack will let steam escape, keeping the necessary pressure from building. Popcorn kernels with the right amount of moisture and unblemished hulls pop into the snack that just about everyone enjoys.

Language Production and Speech Errors

Cognitive psychologists believe that our language construction is the result of a two-step word-retrieval process. The first step is the search for a word's meaning, and the second step is the search for the actual sound of a word. Different speech errors occur when you get stuck at different stages of this two-step retrieval process.

When you incorrectly substitute one word for another, calling a cat a "dog," for instance, your brain has erred in the first step by choosing an incorrect word for the concept of "small domesticated furry pet."

When you have a word or a name on the tip of your tongue, you've gotten stuck in the middle of the retrieval process. You've correctly retrieved the meaning of the word but have been unable to retrieve the sound of the word. Because you've successfully completed one of these two steps, you can probably give the word's definition but can't actually say the word. Going through the alphabet one letter at a time might help you remember it. When you come to the initial letter of the word, your memory might be jogged enough to help you complete the two-step process and remember the sound of the word you're looking for.

When you use an incorrect word that sounds similar to the word you mean to say, your brain has erred in the second step, when you were looking for the actual sound of the word. So, for instance, if you said "Moment of Silence" instead of "Moment of Science" when you were talking about this book or the radio show, your brain would have successfully completed the conceptual stage of the word-retrieval process but then would have made a slip during the second stage by choosing another, similar-sounding word that was more familiar than the word you needed.

Interrupted Vision

Have you ever noticed that you can't see your eyes move when you look in a mirror? Put your face close to the bathroom mirror and look yourself in the eye. Notice the appearance of your eyes. Now, without blinking, look down at your nose, but continue to notice your eyes. Your eyes look different. They've moved, but the movement itself was invisible—to you.

If you have someone else do the same thing while you watch, you'll easily see the other person's eyes move. To make the comparison as fair as possible, you should put your face beside the other person's and look at the other person's nose in the mirror. You've just witnessed a mysterious process that shuts off our vision, at least partly, whenever our eyeballs move rapidly.

The kind of quick eye motion we're talking about is called a saccade, from a French word meaning to twitch or jerk. During the fraction of a second that a saccade takes, images sweep over our retinas at high speed. Yet we don't get a feeling of motion, because our brain suppresses visual perception during saccades. Otherwise, the world might look to us like a bad home video where the photographer held the "record" button down while swinging the camera around the room.

An odd thing about this suppression is that it's not complete. Get in the car and have someone drive you past a roadside fence. Without moving your head, glance quickly from front to back; you can make the fenceposts seem to freeze for an instant. Why, in this case, is vision not suppressed?

A team of visual scientists published a study of this question in the journal *Nature*. They found that what gets suppressed during a saccade are large areas of light and dark. Those are the perceptions that seem to contribute the most to a sense of motion. Finer details, like fenceposts, are not suppressed, maybe because there's no need to suppress them—rapid eye motions usually turn them into a blur anyway.

Burr, David C.; M. Concetta Morrone; and John Ross. "Selective Suppression of the Magnocellular Visual Pathway during Saccadic Eye Movements," and accompanying "News and Views" article by Michael J. Morgan. *Nature*, October 6, 1994.

Why Teflon® Is Slippery

Teflon is the trademark name for PTFE, a type of plastic. If you own any non-stick cookware, then you probably use PTFE on a daily basis. You might not realize, as you fry your morning eggs, that PTFE is one of the most slippery materials that can be manufactured. It's about as slippery as wet ice. What makes Teflon so slippery?

Teflon is chemically similar to another, more common plastic: polyethylene, the material used to make plastic bags and other plastic containers. Chemically, polyethylene is made from long chains of carbon atoms with hydrogen atoms bonded to the sides of the chains. To make Teflon, the hydrogen atoms of polyethylene are replaced by fluorine atoms.

138   It's the fluorine atoms that give Teflon its slipperiness. Fluorine atoms are physically bigger than hydrogen atoms. Their large size makes them huddle around the central carbon chains in a much tighter formation. This tight formation works like a kind of chemical armor, protecting the carbon atoms which in turn hold the molecule together. This chemical teamwork between carbon and fluorine makes Teflon extremely chemically stable, and it's this chemical stability that makes Teflon so slippery. Foreign substances, like a frying egg, can find no chemical foothold on the fluorine armor, so they simply slide away.

Getting this slippery substance to stick to a frying pan is a bit of a trick. Teflon is broken into a fine powder and suspended in water. The pan is then thoroughly cleaned, then roughened by sandblasting. The Teflon is sprayed onto the pan and baked, causing it to fuse together and lock onto the roughened surface of the pan. As long as you don't scratch this protective coating, years' worth of fried eggs, melted cheese, burned milk—even toffee—will slide away effortlessly.

How in the World? Pleasantville, N.Y.: Reader's Digest Association, 1990.
The New Illustrated Science and Invention Encyclopedia. Westport, Conn.: Stuttman, 1987.
Vergara, William C. *Science in Everyday Life.* New York: Harper and Row, 1980.

Your Genetic Cookbook

To stay healthy, your cells must continuously perform a wide variety of chemical tasks. Your DNA genes, in the nucleus of every cell, help orchestrate this activity. A cell uses its genes as a kind of biochemical cookbook: whenever it has to do anything, a cell looks up the recipe in its DNA. It then uses this recipe to cook up whatever proteins it needs to get the job done.

Your cells are hard at work right now, creating thousands of tailor-made proteins for thousands of different biochemical jobs. Sometimes, however, things can go wrong. When a virus attacks, it injects its own DNA into one of your cells. Like your own genes, the viral DNA contains recipes. Unfortunately, these recipes are good only for making new viruses. No matter what your cell was doing before, it will stop its healthy functioning and make only viruses, following the viral recipe until it runs out of raw materials. Then your once-healthy cell bursts open, releasing new viruses to infect more cells.

Cancer is another problem that can develop through a cell's recipe-reading process. Each cell contains dozens of genes that regulate its healthy growth and reproduction. If these genes become damaged or altered, the results can be dangerous. Instead of growing in a healthy fashion, the cell grows and divides at an alarming rate. A cancerous tumor is made up of cells like these, cells whose growth and reproduction genes are altered.

So viral infections are caused when alien recipes are added to a cell's genetic cookbook, and cancer is caused by errors within your own genetic recipes.

Cohen, Jack S., and Michael E. Hogan. "The New Genetic Medicines." *Scientific American*, December 1994.

Shroyer, Jo Ann. *Quarks, Critters and Chaos.* New York: Prentice Hall, 1993.

What Is Jell-O®?

It's pretty easy to guess what most of your food is made of. The meat in your hamburger probably came from a cow, the bun was made from grain, the ketchup from tomatoes, and so on. There is

139

one kind of food, however, that might have you guessing animal, vegetable, or mineral for the whole meal. But no more, because here's the scoop on Jell-O.

Jell-O is a trademark name for gelatin, a food that people have enjoyed for generations. To answer the question animal, vegetable, or mineral: Gelatin is an animal product. It is prepared by soaking the bones, skin, or connective tissue from pigs or cows in a bath of mild hydrochloric acid solution. After this, the animal products are heated in distilled water for many hours, and finally boiled. The fluid that collects from this process is drawn off. What's left is dried and ground into a fine, pale yellow powder. This is unflavored gelatin, the basis of all gelatin salads, desserts, and drinks. Gelatin is valuable as a food because it is pure protein and it is easy to digest. A gelatin made from vegetable protein, called agar-agar, is made for vegetarians.

When powdered gelatin is stirred into hot water and then chilled, it forms the jiggly, gelled product that we are most familiar with. This gelled product is not strictly a solid or a liquid. Technically it's a colloid: a liquid suspended in a solid framework. As the hot mixture cools off, the long strands of animal protein lock together into a solid framework, trapping tiny droplets of liquid water in the process. It's these droplets of water, trapped in a mesh of animal protein, that hold all the coloring and flavoring in your favorite gelatin dessert.

"Gelatin." In *McGraw-Hill Encyclopedia of Science and Technology,* 6th ed. New York, 1987.
"Gelatin." In *World Book Encyclopedia.* Chicago, 1994.
McGee, Harold. *On Food and Cooking: The Science and Lore of the Kitchen.* New York: Scribner, 1984.

Palm-Reading Scientists

In fairy tales and other stories, people curious about their fate could consult a palm reader who, as the story goes, would tell their fortune by looking at the lines on their palm. While there's no scientific evidence linking the lines on your palm to your fate or fortune, scientists have learned that your palm and fingerprints do have a story to tell. The medical study of palm and fingerprints is called dermatoglyphics.

The ridges on your hands, fingers, feet, and toes formed while you were still in the womb, five or six months before you were born. They are the result of stress patterns that formed as your hands and feet developed. Because the growth pattern of every fetus is slightly different, your finger and palm prints are absolutely unique. Even identical twins have slightly different patterns.

Differences in fingerprints have always been useful to police detectives, but what can a doctor tell from looking at your prints? Actually, quite a bit. Many genetic diseases affect the way that the fetus develops. This results in characteristic irregularities in the palm prints. Scientists have statistically linked dozens of genetic diseases to unusual palm prints.

Sometimes even viral diseases can leave telltale traces on an infant's palms. For example, women who caught German measles early in pregnancy during the 1960s sometimes passed birth defects along to their children. A study in 1966 found that such children had characteristic palm and fingerprints as well. Studies have linked irregular palm prints to such diseases as schizophrenia, fetal alcohol syndrome, and even allergies. While they can't tell you how long you'll live or how many children you'll have, the lines on your palm can tell you something.

"Integumentary Patterns." In *McGraw-Hill Encyclopedia of Science and Technology,* 6th ed. New York, 1987.

"Integumentary Systems." In *New Britannica,* 15th ed. Chicago, 1991.

Smith, Antony. *The Body.* New York: Viking, 1986.

Various *Medline Express* abstracts, including "Dermatoglyphic Asymmetry in Fetal Alcohol Syndrome," "Genetic Loadings in Schizophrenia: A Dermatoglyphic Study," and "Dermatoglyphics in Nasobronchial Allergic Disorders."

Our Biological Clocks

Each of us has an internal clock that, among other things, dictates when we get sleepy and hungry. Scientists call our biological clocks "circadian clocks." The word "circadian" comes from two Latin words: *circa,* "about," and *dies,* "day." Our internal clocks are almost parallel to the twenty-four-hour cycle of a day, but not quite.

Research has shown that most people's circadian clocks, left on their own, work on an approximately twenty-five-hour cycle. Without any external stimulus, our internal clocks would usually gain about an hour each day and would be synchronized with the earth's time only one day out of every twenty-four. In effect, our biological clocks must reset themselves each day to become attuned to the twenty-four-hour clock we all live by. Scientists are not exactly sure how the resetting of our clocks happens, but they are fairly confident that our brains utilize sunlight to fine-tune our internal clocks.

Our biological clocks influence practically all of our bodily functions. Our temperatures rise and fall according to these clocks. And because our bodies are set to a daily rhythm, we react to chemical and physical stimulus differently at various times of the day. For instance, studies have shown that our livers process alcohol more efficiently in the evenings than in the mornings, and that we are more likely to have allergic reactions in the middle of the night than in the afternoon.

But what would happen if we were isolated from the influences of the sun, clocks, or any other devices that would mark time for us? In an experiment by researcher Rutger Wever, human subjects were placed in isolated rooms for a month. With no windows, clocks, or television sets, these people, each in a separate room, had no way of knowing what time it was.

Wever found that these subjects extended their sleeping and waking cycles each day, without any knowledge of doing so. Some would sleep as long as seventeen hours at once, and then stay awake for as long as thirty hours. Though the sleeping and waking cycles of the subjects extended far beyond the standard twenty-four-hour period, the biorhythms of the subjects' bodies kept approximately a twenty-five-hour internal clock. The subjects' temperatures were routinely monitored, and the rising and falling of these temperatures never varied much from their twenty-five-hour clock. These results showed Wever that the rising and falling of a subject's temperature is dictated by an internal clock and not by sleeping and waking patterns.

At the end of the experiment, subjects often would be astonished to find that the month was over because, according to their own calculations, based on their sleep cycles, they had been in isolation for only two or three weeks. They were convinced that thirty days had passed only after being shown newspapers that appeared to be from their future.

Winfree, Arthur T. *The Timing of Biological Clocks*. New York: Scientific American Books, 1987.

Déjà Vu

"Have you really been there before?"

Many people at one time or another have experienced déjà vu. French for "already seen," déjà vu is a sudden strong feeling that a moment identical to the present one has occurred at some earlier time.

To a cognitive psychologist, déjà vu is proof of the immense amount of knowledge and experience we store in our brains. When we experience déjà vu, what actually happens is that, in a fraction of a second, we retrieve bits of many different memory fragments and piece them together, producing what seems to be a complete memory. So, if you experience déjà vu in a mall restaurant while waiting for a pepperoni pizza with your best friend, your mind has taken perhaps hundreds of stored memories of various experiences, and put together fragments from those memories to give you the sensation of having been there before, even though you haven't been there before at all.

Cognitive psychologists who study how we use language are not surprised at the brain's ability to create déjà vu. Actually, language comprehension and déjà vu have many parallels. When you hear someone speak, you usually understand her even though you've probably never heard her words presented in exactly the same way. You understand these sentences because your brain is able to remember the individual meanings of words, based on hundreds of past experiences with those words. Your brain takes the meanings of individual words and splices them together to comprehend their meaning as a whole. As with déjà vu, this entire process happens in a split second.

Fresh Fruit in January

Ripe fruit spoils fast, and since freezing ruins most fruit, storing and shipping fruit has always been a problem. Not long ago, fresh fruit in the grocery store had to be locally grown—and still in season.

In a book entitled *On Food and Cooking*, Harold McGee offers this history of how supermarkets came to offer fresh fruit in January:

"In the Caribbean islands, around 1910, it was reported that bananas stored near some oranges had ripened earlier than the other bunches. In 1912, California citrus growers noticed that green fruit kept near a kerosene stove changed color faster than the rest. What secret ripening agent did the stove and the fruit have in common? The answer came two decades later: a simple compound of carbon and hydrogen called 'ethylene,' which is produced naturally by most fruits—and, incidentally, by burning kerosene. The naturally produced ethylene stimulates the processes that we know as ripening: namely, a softening, a sweetening, and a change in color."

With a knowledge of ethylene, the fruit companies could now pick unripe fruit for packing and shipping. The firm, unripe fruit is less likely to spoil or bruise than ripe fruit. When it arrives at the supermarket, the crates of unripe fruit are then gassed with ethylene to prepare them for your kitchen. Ethylene also helps us get fruits that are out of season, since unripe fruit can be stored a lot longer than ripe fruit. Months later, a dose of ethylene restarts the ripening process.

Ethylene doesn't completely duplicate nature, and most fruits still taste best when they're allowed to ripen on the tree. But as long as there's a market for exotic fruits, the fruit industry can use ethylene to satisfy our out-of-season tastes.

McGee, Harold. *On Food and Cooking: The Science and Lore of the Kitchen.* New York: Scribner, 1984.

Ray, Peter. *Botany.* Philadelphia: Saunders College Publishing, 1983.

Why Honey Turns Hard

If a little honey on a piece of homemade toast or in a cup of tea is how you like to start the day, you're not alone. Cave paintings show people a thousand years ago enjoying honey. One drawback to honey, though, is that after sitting too long on the shelf it crystallizes, and that soft, amber liquid turns to a hard, gooey mass.

Actually, though, only part of the honey is crystallizing. Honey is made mostly of two kinds of sugar: glucose and fructose. What crystallizes is the glucose, so the more glucose there is in comparison to fructose, the more likely it is to crystallize. Some honeys, like those made from the nectar of tupelo, locust, or sage, contain slightly more fructose than glucose and so they crystallize more slowly.

But before honey can crystallize, it needs what's called a "seed" for the crystals to grow on. The seed might be a grain of pollen, a speck of dust, or even a scratch on the inside of the jar. But the best seed of all is a bit of honey that has already crystallized. Most of the honey in a supermarket has been heated and filtered to remove virtually all the possible seeds. That slows the crystallization, but the heating process also drives off some of the honey's distinctive flavor. When honey does crystallize, you can soften it again in a microwave or a pan of warm water, but as it cools, the crystallization will begin again—faster even than before.

Honey crystallizes faster the second time because heat alone can't remove all the seeds. Dust, crumbs, and other tiny particles that have accumulated since you first opened the jar will remain as seeds to start the process all over again.

McGee, Harold. *On Food and Cooking: The Science and Lore of the Kitchen.* New York: Scribner, 1984.

Food and Mood

When we think about a pill that helps an overweight person take off the pounds, we usually think of an appetite suppressant. But suppressing your appetite will help you lose weight only if the reason you eat is that you are hungry. Some people eat because they are depressed rather than because they're hungry. When these people want to lose weight, doctors may prescribe not traditional diet pills, but antidepressants.

These antidepressants have helped many overweight people stop craving high-calorie carbohydrates. Depressed people often crave carbohydrates because eating carbohydrate-rich food lifts their mood. As a side effect, they gain weight. Though scientists are not exactly sure how carbohydrate intake influences mood, they do know that it has to do with the chemical messengers, called neurotransmitters, that allow two areas of the brain to communicate: the area responsible for appetite and the area responsible for mood.

More specifically, an intake of carbohydrates increases the levels of the neurotransmitter responsible for mood, called serotonin. Though only a small percentage of the brain's neurotransmitters are made up of serotonin, if this small amount is even minutely tinkered with, a patient can experience fairly drastic mood shifts.

A patient taking antidepressants may stop craving carbohydrates because these drugs function like carbohydrates. They alter the serotonin that communicates messages between the mood and appetite centers of the brain. Once an antidepressant drug begins to work, the abnormal craving for carbohydrates is interrupted and the patient can start losing weight.

Number Crunching at the Electronic Feast

Bits, bytes, kilobytes, megabytes: Such terminology might have you wondering whether you're in a computer store or a fast-food restaurant. Maybe this Moment of Science will make computer bytes a little easier to digest.

Bits and bytes measure the amount of information that a computer's memory or disk drive can hold. Bits are the smallest of these. Computers can process information only as ones and zeros, and a single bit holds one of these—a one or a zero—nothing else. It's like a light switch, either on or off.

A byte is eight computer bits grouped together. You can think of it as a set of eight light switches in a row. While a single switch can have only two positions, on or off, a set of eight switches can be turned on or off in a variety of patterns. As it turns out, there are 256 unique ways that eight switches can be set. Therefore, although it's made of only ones and zeros, a byte can hold any number from 1 to 256.

Grouping bits into bytes makes it possible for computers to handle non-numeric information such as text. Most computers have a built-in table that assigns each of the 256 values a byte can hold to a unique letter, numeral, punctuation mark, or graphic symbol. Working from this table, the computer can use one byte of its memory to store each character in a text document.

A kilobyte is roughly a thousand bytes, and a megabyte is roughly a million. The newest measure—a gigabyte—is a billion bytes' worth of information. A modern one-gigabyte hard drive can hold more text than all the volumes of an encyclopedia. Our number-crunching appetite keeps growing, though: You need four gigabytes to hold all the sound and video from a Hollywood movie.

Measuring Earthquakes

You feel the earth move under your feet: You're in an earthquake. If you were a seismologist, you might be thinking about the best way to measure this unruly natural phenomenon. That was the feeling of Charles Richter, who in 1935 devised a new method for measuring the strength of earthquakes that plagued his native Southern California. Unlike previous scales, which just estimated earthquake damage, his used measurements from a specific type of seismograph. This was one of the first scales to attempt to measure an earthquake's actual intensity.

Although you're likely to hear the term "Richter scale" still used in current news stories, most modern seismologists have replaced the original Richter scale with more sophisticated re-finements. We've come to realize that not all earthquakes shake in quite the same way. While the original Richter scale may have worked well for a Southern California–type quake, it doesn't let us compare them to other kinds of quakes.

Today's seismologists look at many different factors of an earthquake, using separate scales to measure different types of seismic waves. Some seismic waves ripple along the surface of the earth, causing the ground to rise and fall much like ripples spreading out on a pond. To measure this type of wave, seismologists use a "surface-wave" magnitude scale.

Deep earthquakes don't make many surface waves, however, but send their energy like a giant shock wave through the earth's interior. A "body-wave" magnitude scale measures these quakes. Even the surface-wave and body-wave scales fail to account for all the energy released by the largest quakes. For this the "moment magnitude" scale is used. Because there are many ways to measure earthquakes, you will sometimes see different amounts reported for the magnitude of the same quake.

But regardless of the scale that is being used to measure the quake, what do the numbers mean? Perhaps there was a magnitude 5 quake, or a magnitude 6 or 7. We can see that the higher the number, the more damaging the earthquake was, but unless you were standing right there, it might be hard to picture what these numbers actually mean. Earthquake strengths vary enormously: The smallest measurable quake releases no more energy than a falling stone, while the largest ones are more powerful than many nuclear bombs. This tremendous range makes it difficult to deal with earthquake numbers directly, so seismologists use a logarithmic scale instead. In a logarithmic scale, each number represents 10 times the amount of the number below it. This means that a magnitude 5 quake has seismic waves that are 10 times larger than those of a magnitude 4. Magnitude 6 waves are a hundred times bigger than magnitude 4 waves.

What's more, the destructive energy released by an earthquake increases in even larger jumps than the wave sizes. For each step you climb on the scale, the amount of energy increases more than 30 times. Thus, although the waves are a hundred times larger between a magnitude 4 and a magnitude 6 quake, a thousand times more energy is released.

Just how much energy are we talking about, anyway? A magnitude 2 quake, which you probably wouldn't even feel, releases the energy of an average lightning bolt. A magnitude 6 quake, which could destroy some buildings, has the same energy as the atomic bomb dropped on Hiroshima. An 8.3 quake, like the one that rocked San Francisco in 1906, was stronger than more than a thousand of those bombs. It released more energy than the Mount St. Helens eruption of 1980.

The Handy Science Answer Book: The Carnegie Library of Pittsburgh. Detroit: Visible Ink Press, 1994.
Monastersky, Richard. "Abandoning Richter." *Science News,* October 15, 1994.
Vergara, William C. *Science in Everyday Life.* New York: Harper and Row, 1980.

Money Changers

Ever wonder how a money-changer machine knows if you've given it a one-, a five-, or a ten-dollar bill? When a bill is inserted into a money-changer machine, it disrupts a light beam from within the machine. This action triggers the motor to pull the bill into the money changer. The machine then begins a procedure by which it first makes sure that the bill is actual currency, and then determines the denomination of the bill.

With a computer chip and measuring devices, the money changer checks the length, width, and thickness of the bill. If the bill is not the exact length and thickness it should be, the changer will reject it and refuse to give you any change. The sensors that evaluate the bill are so sensitive that even an old, wrinkled bill usually will not pass this authenticity test because it will not measure precisely the same as a crisp, new bill.

After the machine measures the bill's width, length, and thickness, it optically scans the bill to determine if it is a one-, five-, or ten-dollar bill. The machine makes this decision by "reading" how much ink is in different places on the bill. The U.S. Treasury Department uses specially manufactured ink that has unique magnetic properties. The machine's optical scanner measures this magnetic ink. And because a one-dollar bill has a different ink pattern than a five- or ten-dollar bill, the computer inside the machine is able to differentiate between these denominations with a quick scan.

Hot and Cold Chirping Crickets

Imagine that you're camping deep in the woods on a summer night. The sun sinks below the horizon and the temperature begins to fall. You shiver, and as you pull your jacket around your shoulders, you wonder just how cold it really is. Here's a clever way you can find out. All you'll need is a wristwatch, your ears, and a little patience.

Start by listening for a chirping cricket. Male crickets around the world make this characteristic sound by rubbing their wings together. The edge of a male cricket's right wing is covered with little ridges, like a file or a washboard. As he rubs his left wing cover across the uneven right wing, he produces a love song that female crickets find irresistible. Female crickets use this amorous chirping to find their perfect mate, listening with a pair of special ears located below the knees. You can use these same serenades to calculate the temperature.

Most insects tend to be more active when it's hot, and more sluggish in cooler weather. Crickets are no exception to this rule. In cold weather, a cricket will do everything more slowly, and as the temperature drops, so does the tempo of his love song. By measuring how quickly he chirps, you can find out approximately how cold it is.

It's easy to do: Simply count the number of chirps a single cricket makes in a 15-second period, then add 37. Although individual species might be a little faster or slower, this should give a fair approximation of the temperature in degrees Fahrenheit. Even if the chirping keeps you awake all night, at least you'll know the temperature.

Hanson, Jeanne K., and Deane Morrison. *Of Kinkajous, Capybaras, Horned Beetles, Seladangs, and the Oddest and Most Wonderful Mammals, Insects, Birds, and Plants of Our World.* New York: Harper Collins, 1991.

150

When a Boy's Voice Changes

For adolescent boys, a changing voice that cracks in the middle of a sentence can be a great embarrassment. Though embarrassing, a cracking voice is a natural part of adolescent development. As a boy goes through adolescence, his secondary sex characteristics develop. One of these characteristics is the rapid growth of the larynx and vocal cords. A boy's voice deepens as his larynx develops because the bigger the vocal cords, the deeper the voice.

In fact, vocal cords are similar to other musical instruments in this regard. The longer the harp string, for instance, the lower the note it plays. Similarly, if you are blowing into a bottle to create a certain pitch, the larger the bottle, the lower the pitch. When it comes to voices, the bigger the vocal cords, the lower they resonate, and the deeper the voice will be.

But why does a boy's changing voice break and crack? For the same reason growing adolescents are often gangly and awkward—because the brain is becoming accustomed to working with bigger body parts. Even for an adult, a consistent and even voice depends on the brain's ability to constantly monitor the sounds that come from the voice. The brain can do this quite easily under normal circumstances. But when a boy's vocal cords enlarge, the brain must relearn how to monitor and control the voice. A cracking voice is proof that an adolescent boy's brain hasn't become completely proficient at coordinating its careful monitoring of the sounds coming from the vocal cords.

How Dogs Eat

If you've ever watched a dog eat, you've probably marveled at how quickly it gulps down its food. You might even wonder why, no matter how hungry a dog is, it will often eat as much food as you put in front of it.

Dog owners may be concerned about this behavior, but it poses no problems for the dog. People chew their food and try to teach their children to eat slowly because digestion for humans begins in the mouth. Our saliva mixes with food and prepares that food to be broken down into its primary nutrients once it enters the stomach. A dog's digestion, on the other hand, doesn't begin until the food reaches the stomach, so dogs do not need to take time chewing their dinners.

Most dogs probably eat so quickly because in the days before they were domesticated, they had to survive by eating their prey before another dog or scavenger animal stole it. The evolutionary programming of dogs dictates that they eat and keep moving. As a species in the wild, they didn't have the luxury of hanging around and eating at their leisure.

Even their teeth aren't designed for them to savor their food. While most of the teeth in a human's mouth are flat and designed to facilitate chewing, most of the teeth in a dog's mouth are pointed and designed to allow the dog to grab its food and swallow it whole.

Hundreds of years of domestication hasn't changed most dogs' eating habits very much. Even if a dog has been given regular, dependable meals every day, it will still gulp those meals down in a flash, ensuring that no scavenger will take its food away.

Johnson, Norman H. *The Complete Puppy and Dog Book.* New York: Atheneum, 1977.

Morning Breath

Morning breath is a topic of many jokes, and a real boon to the mouthwash and toothpaste industries. But what exactly causes the foul taste and odor in your mouth when you wake up in the mornings?

The most basic answer to this question is gravity. If you've eaten fewer than three hours before you go to sleep, your stomach hasn't had time to digest all of your food. When you lie down, gravity causes the gastric juices that are busy digesting your food to come up into your throat.

Because your airway and food pipe are side by side, they share an adjoining wall. Those gastric juices that have backed up in your food pipe actually permeate that wall and enter into your air pipe. These juices can irritate your larynx, and cause your voice to be hoarse and your breath to turn foul. In addition, gastric juices are acidic enough to burn the mucous membranes in your throat, resulting in a sore throat that occurs in the mornings but lessens throughout the day.

Morning mouth tends to worsen with age, because as you get older, the top of your esophagus, the thin tube that leads from the mouth to the stomach, loosens. With the softening of the top of this tube, gastric juices can more easily escape from the esophagus into the air pipe.

Another contributing factor to morning breath is infrequent swallowing during sleep. Swallowing allows us to keep our mouth relatively free from odor-producing bacteria. But when we're asleep, these bacteria can thrive, contributing to the bad taste in our mouths when we wake up.

Lowering the Thermostat

When you lower your thermostat at night, are you really saving energy? Or does it take *more* energy to *reheat* a cold house? This Moment of Science will tell you why turning *down* your thermostat will never *raise* your heating bills.

To see why, it may help to think of the house as a bucket with a hole in the bottom, and heat as water being poured into the bucket. The pressure of the water in the bucket forces water out the hole— the more water there is in the bucket, the greater the pressure, and the faster the water will leak out. But as the water level drops, the pressure will drop too, and the leak will slow down. So it would take more water to keep the bucket full than it would to let some or even all of the water leak out and then refill it.

When it's cold outside and warm inside, that temperature difference acts like the water pressure in the bucket, forcing heat out through the walls. Heat leaks more slowly out of a cooler house because there is less heat trying to get out. It may take a lot of energy to heat a cold house, but not nearly as much energy as would have been saved by letting the temperature drop.

Some houses with what's called a high "thermal mass" keep a steadier temperature by absorbing heat during the day and giving it off slowly at night. In these houses a lower thermostat won't save as much energy, but it'll still save some. That's because thermal mass is like a sponge in the bottom of the bucket: It may slow the leak, but it can't reverse the process.

Turning the heat down too far could get uncomfortable or even cause the pipes to freeze. But it won't raise your heating bill.

Learning to Catch

It's a line drive headed straight for your glove. All you have to do is close your hand at the right moment. It takes only about fourteen one-hundredths of a second to close your hand around a baseball, but in that time a baseball going 90 miles an hour travels more than 18 feet. Closing your hand a little late or a little early could turn the final out into a home run.

But how do we know when the ball is exactly fourteen one-hundredths of a second away? We could use two variables—the ball's speed and its distance—to calculate when it would arrive. But our subconscious mind does it without calculating and with only one variable. That variable is the rate at which the image of the ball is expanding in relation to the size of the image at that moment.

Here's how it works: A ball thrown from a distance grows gradually from a tiny speck until it reaches us. The speck grows slowly at first, but the closer it gets, the faster it grows. A big ball far away may look the same size as a small ball up close, but if they're traveling at the same speed, the closer ball grows faster.

Knowing how big the speck is and how fast it's growing, you could calculate how long it will take to reach you, but our subconscious mind does it in a single glance. By interpreting a single variable—that is, how fast the ball appears to be growing relative to how big it appears at that moment—our mind knows just how long it will be before that speck turns into a baseball in our hand.

Savelsbergh, G. J. P.; H. T. A. Whiting; and R. J. Bootsma. "Grasping Tau." *Journal of Experimental Psychology: Human Perception and Performance* 17, no. 2 (1991).

Biodiversity and Genetic Engineering

If you want a single yardstick to measure the overall health of Planet Earth, most biologists would say that this yardstick is biodiversity. Biodiversity is a measure of the number of different species that exist. For example, a small forest with 50 different kinds of trees has more biodiversity than a large plantation with only 1. Why is biodiversity so crucial to the health of earth's ecosystem?

The more species earth has, the better able it is to survive—and stabilize—a changing climate. You can think of earth's ecosystem as a large building, supported by many beams and pillars. As different species become extinct, it's like knocking out a beam here, a pillar there. The building might not collapse at once, but the next change of wind could spell disaster.

You might think that genetic engineering, which develops new strains of plant and animal species, would help to increase biodiversity. It can, but unfortunately there are also ways that genetic engineering might reduce biodiversity rather than aid it.

One way genetic engineering might harm biodiversity is by encouraging irresponsible farming techniques. For example, if scientists developed a strain of corn that's cheaper and easier to grow, farmers might plant this variety only, letting others fall by the wayside. This is called monocultural farming, and it's extremely risky. If a plant disease came along that attacked this strain of corn, we wouldn't have others to fall back on. This happened on a small scale in 1970, when a blight wiped out 15 percent of the U.S. corn crop. If genetic engineering encourages more monocultural farming, then the 1970 blight could be just the tip of the iceberg. We can only hope that genetic engineering will make us more sensitive to the value of our planet's biodiversity. Our future depends on it.

Doyle, Jack. *Altered Harvest.* New York: Viking Press, 1985.

Fox, Michael W. *Superpigs and Wondercorn.* New York: Lyons and Burford, 1992.

"Genetic Engineering." In *Opposing Viewpoints Sources: Science and Technology.* San Diego: Greenhaven Press, 1990.

Leary, Warren E. "F.D.A. Approves Altered Tomato That Will Remain Fresh Longer." *New York Times,* May 19, 1994.

The Secret Life of Hiccups

We've all experienced them at one time or another, often right after a big meal. A normal bout of hiccups usually lasts only a few minutes and may contain up to 70 individual "hics." Unlike coughing or sneezing, which can help clear your airways, hiccups seem to serve no beneficial function in the human body. What's the story behind these strange convulsions?

Two separate things happen to your body when you hiccup. The muscles in your diaphragm, which normally control your breathing, contract with a sudden jerk. This causes a sharp intake of breath. At the same time, your vocal cords contract to stop this breath, resulting in a loud "hic."

This is all caused by a misfire in the nerves that control your diaphragm. These nerves run from your neck to your chest, and any unusual pressure or irritation along this length can cause a misfire. Thus, hiccups are often triggered by overeating, gulping your food too quickly, or eating something too hot or too cold. Stress can also cause hiccups.

There are many folk remedies for hiccups, but none seems to work for everyone. Such remedies include holding your breath, breathing into a paper bag, or drinking a glass of water without breathing. It's possible that by depriving the diaphragm muscles of oxygen, these remedies force them to resume a more normal breathing pattern.

Other remedies include pulling your tongue, sucking a lemon, or having a friend startle you. What these remedies have in common is that they trick your nervous system with a diversion, perhaps shocking the nerves that control the diaphragm into normal behavior. No one knows exactly why these remedies sometimes work.

It's extremely rare, but severe cases of hiccups do occur. If you have persistent hiccups that simply refuse to go away, you should probably consult a physician.

"Curing Hiccups." University of California, Berkeley Wellness Center, October 1985.
"Hiccups." In *ABC's of the Human Body,* ed. Alma Guinness. New York: Reader's Digest Press, 1987.

"Hiccups." In *Mayo Clinic Family Health Book,* ed. David E. Larson. New York: Morrow, 1990.

"Hiccups." In *Prevention's Giant Book of Health Facts,* ed. John Feltmann. Emmaus, Pa.: Rodale Press, 1991.

Noise from the Upstairs Neighbors

Are you annoyed by thuds and thumps when your upstairs neighbors walk around? And if so, what can you do about it? A surprisingly discouraging answer to that question appeared in the *Journal of the Acoustical Society of America.* After investigating the question, a private consultant and a California state noise-control officer report that "in multifamily, wood-frame residential construction . . . at present, there is no economically practical method of avoiding the perception of 'thuds' and 'thumps' in rooms beneath the walking surface."

157

The occasion for this study was a lawsuit. Homeowners on the lower floors of a Northern California condominium complex sued the builder for $80 million because of thumps, booming, and other "feelable" structural vibration as upstairs neighbors walked around. The $750,000 condos had been marketed as having "luxury" acoustical privacy.

In a quiet empty warehouse, the researchers built a full-size mock-up of a pair of stacked rooms like those in the condo in question. Then they studied the sounds in the lower room caused by a standard tapping machine and a real person walking on the upper floor.

They found that resilient mats, carpeting, and the use of sneakers or bare feet upstairs would eliminate high-frequency noises such as clicks and scrapes. But the low-frequency sounds—the thuds and thumps—were caused by the floor vibrating like a giant drumhead. Track shoes actually made those sounds worse. And resilient mats upstairs seemed to encourage the walker to move with a springier gait, making the booms below louder still. In short, these two researchers found no feasible way to make the upstairs neighbors inaudible in a wooden residential building.

Note: The suit was settled out of court in favor of the plaintiffs—but for a lot less than $80 million.

Blazier, Warren E., Jr., and Russell B. DuPree. "Investigation of Low-Frequency Footfall Noise in Wood-Frame, Multifamily Building Construction." *Journal of the Acoustical Society of America,* September 1994.

Bright Light Can Help You Sleep

Most people's sleep schedule follows day and night cycles. We are awake during the day when it's light, and asleep during the night when it's dark. This cycle is controlled by an internal biological clock that tells our bodies when to get sleepy and when to wake up. Research has shown that this internal clock works even if a person is put in a cave that never receives any light. So, exposure to the sun's light doesn't establish our internal clocks, but it does help fine-tune them. In fact, light plays a very important role in helping us sleep. Exposure to light during the day reinforces our internal clocks by making us feel more sleepy at night and less sleepy in the day.

This happens because retinal fibers in our eyes are connected to the part of our brain that controls our internal clock. When those fibers sense bright light at midday, they send a message to the brain that it should get sleepy later in the day. In theory, the more exposure to bright sun you get during the day, the stronger this message will be and the better you will sleep at night.

This works well for those who work during the day and sleep at night, but what about night workers? If you are up at night and asleep during the day, you are fighting the influence of light on your biological clock. However, you can work with light to make things easier by tricking your body into believing that night is day and day is night. To do this, increase the brightness of the lights you work around and make the room you sleep in as dark as possible. By tricking your internal clock, you will be more likely to stay awake at work and then go home to get a good day's sleep.

Kelly, Dennis D. "Sleep and Dreaming." In *Principles of Neural Science*, 3rd ed. New York: Elsevier, 1991.
"Setting the Body's Clock for Sleep." *Consumer Reports on Health*, March 1994.

Why Ice Is Not Slippery

Try telling someone who has just fallen on a patch of ice that ice is not slippery and they'll think you're crazy. But, in fact, ice itself isn't slippery because it is a solid. One quality of solids is that when two solids are together, there is friction between them that will keep them from slipping.

So how can your shoe slip on ice? The answer lies in two peculiar properties of ice. The first is that as water freezes, its molecules move farther apart. The molecules of most substances move closer together as they freeze, making them shrink at lower temperatures. But water molecules move farther apart at temperatures below 39 degrees, making water expand as it freezes. That is why frozen water pipes burst, and a tray of ice cubes will freeze over its top if you fill it too full.

The second peculiar property of ice is directly linked to its first peculiarity. When subjected to pressure, ice melts. Remember that the molecules in ice are farther apart than the molecules in water; therefore ice molecules are vulnerable to pressure which pushes them closer together, causing the ice to change into water. 159

So when you step on a patch of ice, you exert pressure on the ice, which causes its molecules to move closer together. That makes them revert to their more dense state, which is water. If you slip on a patch of ice, then, you in fact are slipping on a thin layer of water that the pressure from your weight has created. And unlike solid ice, water as a liquid is quite slippery.

Cazeau, Charles C. *Science Trivia: From Anteaters to Zeppelins*. New York: Plenum Press, 1986.

Flatow, Ira. *Rainbows, Curve Balls and Other Wonders of the Natural World Explained.* New York: William Morrow and Company, 1988.

McGee, Harold. *On Food and Cooking: The Science and Lore of the Kitchen*. New York: Scribner, 1984.

How Fruits and Vegetables Help Prevent Cancer

Everyone has heard that fruits and vegetables are good for you, but scientist Gladys Block has proven it as a fact. Block took the results of 156 separate scientific studies and determined that, except for quitting smoking, the best way a person can help prevent cancer is to eat more fruits and vegetables. According to these studies, people who eat the most fruits and vegetables have half the cancer risk of those who eat the least.

One of the major reasons fruits and vegetables are such great cancer fighters is that they help combat the ill effects of oxygen circulating through our bloodstreams. Of course, oxygen helps keep us alive, but it also takes a toll in the process. As oxygen travels through your bloodstream, some oxygen molecules lose an electron. Because electrons usually travel in pairs, these unpaired electrons, called free radicals, are chemically unstable. To make themselves stable again, free radicals take electrons from other molecules in the body, which damages these molecules. Over time, the damage caused by this electron thievery can alter a cell's DNA and can eventually lead to cancer.

In order to combat the ill effects of unstable oxygen molecules, you can eat fruits and vegetables high in vitamins A and C. These vitamins donate electrons to the free radicals and, in the process, stabilize them and keep them from needing to steal electrons from molecules in the body. In the long run, this can keep the body's cells cancer-free.

As a rule of thumb, when choosing fruits and vegetables, go for those that have the brightest colors. Spinach, carrots, and other brightly colored vegetables usually have lots of vitamins A and C. These foods are so good at stabilizing free radicals that the National Cancer Institute recommends that we eat at least five servings of them a day.

Castleman, Michael. "Mother Knew Best." *Sierra,* December 1994.

Honey from Honeydew

We've all heard that bees make honey, and for the most part that is where honey comes from: The bees extract nectar from flowers, which they take back to their hives, to be made into honey. But in areas without flowers or in times when flowers are scarce, another insect joins the honey-making process. Aphids suck the sap out of the stems of plants, eating some, but letting a lot more drip onto the plant or the ground.

If you've ever parked your car under a tree filled with aphids, you may have come out to find it covered with tiny sticky spots. Those spots are the sap sucked out of the tree and dropped by the aphids. That sap is called "honeydew" because it looks like a sweet dew. Honeydew makes a mess of your car, but for hungry bees, it's a reasonable substitute for nectar. Bees collect it, just as they collect nectar from flowers, and bring it back to the hive to be made into honey.

Honey made largely from honeydew doesn't taste as good as honey made from nectar because it contains a large number of bad-tasting and undigestible proteins. Even bees restricted to a diet of honeydew honey may sicken and die. But when there's no nectar to be had, honeydew—extracted by aphids from the stems of plants—may provide at least a temporary substitute and a way to supplement a hive's supply of honey.

In fact, bees will convert anything with a high enough sugar content to honey. Urban beekeepers even report that after a large outdoor concert in a nearby park, bees that have been feasting on the dregs of soft-drink containers produce honey with a similar flavor.

McGee, Harold. *On Food and Cooking: The Science and Lore of the Kitchen.* New York: Scribner, 1984.

Measuring the Pressure in Your Eyeball

Often when you go to an optometrist to have your eyes checked, you have to sit in front of an instrument that shoots a puff of air into your eye. The purpose of that puff of air is to measure the pressure in your eyeball.

Your eyeball has pressure that can be measured because it is filled with fluid, just like a water balloon. The more water in a balloon, the more pressure there will be on the wall of the balloon, and the more potential there is that the balloon will be damaged.

The same is true for the eyeball. If fluids build up in an eyeball, pressure will increase on the tissues of the eye. In an unhealthy eyeball, this pressure can increase to the point of damaging the nerve fibers. If nerve fibers are damaged, they cannot get information back to the brain, and vision becomes impaired.

So how does that special instrument measure the pressure in your eyeball? By measuring the shape of the eyeball during the test. When the puff of air hits the eye, it momentarily flattens out a small part of the cornea, the clear, curved surface on the front of the eye. The instrument measures the amount of time it takes for the puff of air to flatten the cornea. An eyeball that has more fluid in it, and therefore has greater pressure, will take longer to flatten out than one that has less fluid.

Think of the water balloon again. If you are holding a water balloon that is as full of water as possible, you won't be able to push on it and change its shape very easily. If the balloon is only half full, any pressure you put on it will change its shape quite quickly.

The information about the pressure in an eyeball can help eye doctors diagnose early stages of eye diseases, especially glaucoma. With regular visits to your eye doctor, this and other eye diseases can be caught early and treated.

162 *Why Tires Have Treads*

If you've ever driven on a highway during a heavy rain, you know about the danger of skidding. This "waterskiing" on the road is technically called hydroplaning, and it happens when water comes between the road and your tires, causing you to lose traction and go out of control. Your car is less likely to slide around when the road is dry because there is enough friction between your tires and the road to keep you steady, even at high speeds. But when it rains, a layer of water builds up between your tires and the road. This water interferes with the friction that helps your tires grip the road surface.

This is where the treads on your tires come in. If the road you are driving along is covered with water, the pressure of the tire against the road surface causes the water to be squeezed up into the tire treads. These treads help your tires pump water out from underneath the tire so that the rubber can be in contact with the road, thus creating the friction that will stabilize your car.

The faster you go, the more water your tires have to remove. That's why your car may not hydroplane at 30 miles per hour, but might at 60 miles per hour. In fact, at highway speeds, during hard rains, each tire must pump away about a gallon of water every second. This is hard for tires to do if they are old and have treads that are worn down. When there isn't enough space in the tire's grooves, water can't find a place to go, and so it creates a layer of lubrication, causing you to do the equivalent to waterskiing on the highway.

"Grabbing the Wettest of Roads." *Discover*, October 1992.

Another Kind of Herpes

You might call them cold sores or canker sores or maybe fever blisters. Whatever they're called in your house, if you get them, you're in the company of about 95 million other Americans who suffer from the virus that causes cold sores. The scientific name for this virus is herpes simplex, type one. Although it is a relative of herpes simplex two, more commonly known as genital herpes, its effects are usually less serious.

As soon as this herpes virus begins producing a cold sore, the body's immune system manufactures antibodies and white blood cells to combat the virus. The immune system may be able to win the battle against an individual cold sore, but it can never win the war against the virus that causes the cold sore. Even after the cold sore is completely healed, the virus remains hiding in the body in a resting, or latent, phase. If triggered, this latent virus will begin the cycle again and cause another outbreak.

In fact, recurrence of these painful cold sores is very common because once you have the virus, you never get rid of it. The latent virus can be triggered by many factors. Exposure to wind and sun can cause an outbreak, and so can physical or emotional stress. Many people who have the virus get a blister every time they come down with a cold because their immune systems get overloaded and can't attend to everything at once.

Some people never get cold sores because they have never been infected with the virus, which you can get through direct contact with the lips or mouth of a person with an active sore. Researchers believe that most people were infected as children when an adult with a cold sore kissed them.

Dorland's Illustrated Medical Dictionary. 27th ed. Philadelphia: Sanders, 1988.
The Merck Manual. 14th ed. Rahway, N.J.: Merck, Sharpe and Dohme Laboratories, 1982.
Stedman's Medical Dictionary. 25th ed. New York: Macmillan, 1990.

Why Kids Can Sleep through Just about Anything

Have you ever been to a party and seen a child sleeping happily on a couch, undisturbed by the adults talking and laughing in the room? How can kids sleep so soundly when exposed to noise and commotion? The answer lies in the difference between how adults sleep and how children sleep.

There are four different stages of sleep, and all sleepers cycle between these stages up to six times each night. The biggest difference between how adults and children sleep occurs in stage-four sleep, which is called slow-wave sleep because in this stage your heart rate and your blood pressure decrease; your brain is less active than at any other time; and your dreams, if you have any, tend to be vague and abstract. Slow-wave sleep is also called deep sleep. Children spend far more time in this stage of sleep than adults.

During deep sleep, you sleep so heavily that you lose control of many of your muscles. Your mouth can drop open and you might drool. You probably won't be awakened by noises and activity around you. For the most part, as people grow older, they spend less and less sleep time in deep sleep. By the age of 60, most people will spend almost no time in this stage of sleep. Children, however, spend most of their sleep time in deep sleep, so those children sacked out on a couch at a party are probably not going to be disturbed, even by a bunch of adults standing around having a good time.

Fighting AIDS, Fighting Evolution

One way our bodies resist disease is by producing a variety of chemicals called "antibodies," which destroy the virus or bacterium causing the disease. Antibodies have to be selective to avoid destroying the wrong cells, so each type of antibody recognizes and attacks only a specific virus or bacterium. And that means that each new disease requires a new type of antibody.

But in fighting disease, evolution sometimes works against us because the faster an organism multiplies, the faster it adapts to changes in its environment. For bacteria, the appearance of a new antibody means a big change in their environment, but unfortunately it's a change the bacteria often adapt to. Since a single bacterium can multiply a thousandfold in three hours, there's a good chance that some of those offspring will vary from the rest in ways that make them unrecognizable to the antibody. Those bacteria will then produce millions more like themselves.

Some viruses, including HIV, which is responsible for AIDS, evolve even faster than bacteria. Unlike bacteria (as well as plants and animals), which carry their genetic information in the form of DNA, these viruses carry their genetic information in a similar molecule called RNA. When RNA is copied to make a new virus, the process is less accurate, and far more mutations occur than when DNA is copied. That means more variety among the offspring and more chances that a few of them will escape the antibodies. In fact, viruses evolve so quickly that the same virus in two individuals often evolves into two different strains. All living organisms adapt to changes in their environment, but RNA viruses, like HIV, adapt much faster because they produce more offspring with more variation.

Losing Your Voice to an Echo

Shout "HELLO" between two cliffs and you'll hear the echoes as the sound waves bounce back and forth. Two things happen as the echoes bounce between the cliffs. First, the echoes get fainter because when the sound hits the cliff, not all of it comes back: some is absorbed by the rock, some bounces off in other directions, and some misses the cliff altogether.

But your voice also gets distorted so that each succeeding echo sounds less like the original sound. Even though your voice contains a wide range of frequencies, what happens to those frequencies depends on the size, position, and material of the objects they come into contact with as they echo between the cliffs. Every situation is slightly different, but inevitably some frequencies are more likely to be reflected while others simply fade away. So after a few echoes, the sound loses the characteristics of your voice and takes on new characteristics unique to the area you're standing in.

Real echoes usually fade away before they become that distorted, but you can see how it works with an artificial echo on two tape recorders. Say "hello" followed by your name into one tape recorder. Then play it back into the microphone of the other one. Play that recording back into the first tape recorder and so on—back and forth about five to ten times. Gradually the two words will sound the same, as they lose their original characteristics and conform to the acoustic characteristics of the room and of the two tape recorders.

Incidentally, this is a problem in concert halls, which have to be specially designed to reflect all the different frequencies accurately.

Sonic Booms

On October 14, 1947, U.S. Air Force pilot Chuck Yeager became the first person to fly a plane faster than the speed of sound. As with every supersonic flight since, Yeager's plane created a loud noise, called a sonic boom, that could be heard on the ground.

When a speedboat moves through water, it creates waves that you can see. Planes create similar waves in the air, which you can hear. As a plane travels, it compresses air molecules in front of it, creating waves that spread in all directions. These waves travel at the speed of sound and are known as sound waves. When a plane itself approaches the speed of sound, which is roughly 750 miles per hour, it begins to move as fast as the sound waves it is creating. When the plane moves faster than its own sound waves, it forces those waves to pile up one on top of another.

This concentration of sound waves becomes what is known as a shock wave. Shock waves are powerful waves that travel in all directions, including toward the ground. At each location where shock waves hit the ground we hear a sonic boom, which is not just a one-time result of breaking the sound barrier. Because the plane is creating shock waves the entire time it flies at supersonic speeds, sonic booms can be heard the entire time as well.

But not all sonic booms are alike. As a general rule, the bigger the aircraft, the larger the shock waves and the longer the duration of the sonic boom. The smallest aircraft capable of traveling faster than the speed of sound generates a boom that lasts one-twentieth of a second. The largest supersonic aircraft creates a boom that lasts ten times longer, or half a second.

However, two aircraft the same size flying at different altitudes will create sonic booms of different intensities. A craft flying higher will create a less intense sonic boom, because the shock waves have farther to travel before hitting the ground, and therefore have more time to dissipate. Although not as intense, the sonic boom created by this high-flying plane will be heard over a larger geographic area than the boom created by a low-flying plane because as the shock waves travel to the ground, they spread out and increase the radius they affect.

Luckily, even the most intense sonic booms can't hurt people, but they have been known to damage plaster walls and break windows.

"Sonic Boom." In *McGraw Hill Encyclopedia of Science and Technology,* 7th ed. New York: McGraw-Hill, 1992.
"Sonic Boom." In *World Book Encyclopedia.* Chicago, 1994.

Spoonerisms

At some point, everyone has transposed the first letters in two words and come up with a nonsense phrase. You might mean to say "barn door," but it comes out "darn boor." These slips of the tongue are called spoonerisms, and cognitive psychologists study them because of what they say about how our brains construct language.

Early twentieth-century psychologists believed that language was produced in our brains one word at a time, that each word acted as a stimulus to produce another word. But cognitive psychologists now believe that we produce language in clumps rather than one word at a time.

The study of spoonerisms has helped scientists formulate these new theories. Spoonerisms may seem like random mistakes, but in fact they follow a regular set of rules. When two sounds are transposed between two words, they are almost always sounds that belong in the same position. For example, the beginning of one word almost never exchanges with the end of another. The close association your brain makes between two words such as "barn" and "door" indicates that your brain chose those words as a unit, rather than one at a time.

We make speech errors like this because as we construct language, our brain builds a frame for what we are going to say before we choose the actual words that will go into that frame. When we get a phrase right, our brains have successfully coordinated this frame with the sound of a word. Spoonerisms happen when this coordination breaks down, often because of the interference of external or internal stimulus.

Viking Tales: Poetic License or Medical History?

In the Icelandic family sagas, composed more than 800 years ago, the intellectual and physical prowess of the Viking poet and warrior named Egil (pron. *Eye*-yihl) stands out as almost beyond belief. According to the saga, Egil's fierce appearance alone was enough to make at least one of his enemies capitulate to the Viking's demands on the spot. And apparently he was as tough as he was fierce: According to the saga, Egil's skull was so strong that after he died, it couldn't be broken even with the full swing of an axe. But if Egil was a thick-skulled warrior, other parts of the saga suggest that he was also a gifted and sensitive poet who suffered from the pain and stigma of disabilities including headaches, blindness, and loss of hearing.

Scholars have often used the axe passage to show how unreliable the Icelandic sagas really are; after all, everyone knows that an axe is tougher than a skull. But a rare disease, named after a nineteenth-century English surgeon, could have made Egil as extraordinary as the sagas claim.

Paget's disease, which produces a thickening and malformation of the bones, also causes blindness, headaches, and loss of balance and hearing—all problems that Egil describes in his own poetry. Even the cold hands and feet Egil describes could have been caused by a loss of circulation associated with Paget's disease. But even with a massively thick skull, Egil could have been a gifted poet, because the bone accumulates only on the outside of the skull, leaving the brain undamaged.

The Icelandic sagas may not be perfect historical records, but the combination of medical and literary evidence suggests that the story of Egil's skull may have been more than a literary device designed to exaggerate the Viking's heroic stature.

Byock, Jesse L. "Egil's Bones." *Scientific American,* January 1995.

Potential Energy

When we think of the concept of energy, we often think of objects in motion: a car rolling down a hill or a jogger running up a hill. This kind of energy-in-motion, called kinetic energy, isn't the only kind of energy. Another type, which motionless objects can have, is called potential energy.

Potential energy is stored energy—that is, energy that can be tapped later. To better understand the concept of potential energy, consider how a crossbow works. If you took a crossbow, pulled its string back, and cocked it so that the arrow remained in place until you decided to shoot it, you would have created energy, energy that can be released by pulling the trigger. You have increased the *potential* energy of the crossbow.

The amount of potential energy in the cocked crossbow is equal to the amount of work you put in it by pulling it back and cocking it. This potential energy becomes kinetic energy when you pull the trigger. With the release of the trigger, the potential energy of the cocked crossbow converts into kinetic energy as the arrow flies through the air. Except for a bit of energy lost to friction, the kinetic energy of the flying arrow is equal to the potential energy that was stored in the cocked crossbow.

Of course, potential energy doesn't exist only in crossbows. Other examples include a boulder on the edge of a cliff and gasoline in the tank of a car. In effect, potential energy is simply kinetic energy waiting to happen.

The Shape of Lightning Bolts

As a thundercloud moves through the air, a strong negative charge gathers near its base. Because opposite charges attract, this negative charge is anxious to combine with the positive charges in the ground. Eventually a lightning bolt forms to neutralize these different charges.

We might think that this bolt would want to jump in a straight line, that the electric charge would try to find the most direct route between thundercloud and ground. Why then are lightning bolts so jagged and irregular?

The answer has to do with the complex way a lightning bolt forms. Although it looks like it forms all at once, a lightning bolt is actually produced in many steps. Instead of jumping right to the ground, the cloud's negative charge begins with a short downward hop. This initial hop is called a leader, and it's no more than a few hundred feet long. From the lower end of this leader, another leader forms, and from the lower end of this, another. In this manner, the negative charge hops downward from leader to leader like a frog jumping from lily pad to lily pad across a pond.

While this is going on, the ground sends up its own chain of shorter, positively charged leaders. It's only when these two chains meet, about a hundred feet off the ground, that we see the lightning bolt's flash.

So lightning is jagged because each leader forms independently of the others. Each place a lightning bolt zigs or zags is where one leader stopped and another one started. Each place a lightning bolt forks is where two separate leaders formed from the bottom end of a single leader above. This whole process takes only a few thousandths of a second, but that's enough time to sculpt beautiful and complex lightning bolts.

Trefil, James S. *Meditations at Sunset: A Scientist Looks at the Sky.* New York: Collier, 1987.

Bugs in Cake Mixes

Those white, wiggly bugs sometimes found in boxes of grain foods go by many names: mealybugs, mealyworms, and, most appropriately, wigglies. These annoying critters aren't precisely bugs or worms, but are the larvae of moths and beetles. Their favorite food is starch, and what better place to find starch than in boxes of flour, sugar, and cake mixes?

Manufacturers of these convenience foods put their products through processes that kill any moth and beetle eggs which might eventually hatch out into the food. So when you find wiggling bug larvae in your pudding mix, you're not looking at the results of a food company's inferior standards.

Most of the time, moth and worm larvae enter a box of food from the outside because they are attracted by the starch in the glue that holds the box together. Once they eat through the box, they find a veritable cafeteria that could feed them for a long, long time.

After eating much of the starch in the foods they have infested, the bug larvae leave the food hard or clumpy, mainly because starch is the part of grain that keeps it from clumping and hardening. Boxes of food that have sat on a shelf for a long time are more likely to have larvae infestations than new food.

The good news about grain larvae is that they are not poisonous in any way. Nutritionists assure us that if most people weren't so squeamish about eating food that moves, they wouldn't need to purge the infested boxes from their shelves. The bad news about these invisible invaders is that if they aren't found early enough, they can ruin an entire pantry of food.

How Time Passes in Dreams

Many people believe that hours' worth of events and activities can be dreamed about in a matter of seconds. Despite this common belief about how we dream, time in dreams actually is not compressed. If you dream of an activity that would take five minutes in waking life, you probably dream about it for a full five minutes.

Dream and sleep researcher William Dement conducted two studies that demonstrated that dream time was similar to real time. Because dreamers' eyes move under their eyelids very rapidly while they are dreaming, Dement was able to monitor sleepers and record the length of their dreams by observing their rapid eye movement.

After recording this information, Dement would wake dreamers and have them write down a description of their most recent dream. He assumed that longer dreams would take more words to describe than shorter ones. When he compared the number of words in each dream report with the number of minutes over which the dream had occurred, he found that the longer the dream, the more words the dreamer used to describe it.

In another related experiment, Dement woke sleepers while they were dreaming and asked them how long they perceived their most recent dream had taken. Eighty-three percent of the time they perceived correctly whether their dreams had been going on for a long time or for a short time. With these experiments, Dement concluded that time in dreams is nearly identical to time in waking life.

So the next time in your dreams you slay a dragon or fly from your house to your workplace, the amount of time it seems to take will probably be just about how long it actually will take to dream it.

Kelly, Dennis D. "Sleep and Dreaming." In *Principles of Neural Science,* 3rd ed. New York: Elsevier, 1991.

The Land of Sweat and Honey

Milk and honey may be emblems of the easy life, but for a bee, making honey is no easy task. In fact, honey making is such hard work that for every pound of honey that goes to market, bees eat eight more pounds just to maintain the activity of the hive.

One bee is so small that it doesn't need much honey to get around: In terms of fuel efficiency, a bee gets about 7 million miles per gallon of honey. But with more than 20,000 bees in a hive, honey consumption adds up fast.

Every day a worker bee makes up to 25 trips between the hive and the flowers to gather nectar, each time returning with a load equal to about half its body weight. But half the body weight of a bee comes to only around two one-thousandths of an ounce. So for all its effort, a bee may work its entire life to make less honey than you stir into a cup of tea.

Back in the hive, the nectar still has to be converted to honey. First, the bees pump the nectar in and out of themselves for 15 to 20 minutes to reduce the water content. At the same time, they mix enzymes into the honey to break the nectar's complex sugar into simpler, more soluble sugars. Next it's spread on the honeycomb, where it evaporates further. To speed the process, the bees continually move the air around by fluttering their wings. In about three weeks, bees pack the finished honey into cells in the honeycomb and seal it with wax.

So relax and enjoy the sweet flavor of fresh honey, but remember that, for the bees, at least, a land of milk and honey would be no easy life.

McGee, Harold. *On Food and Cooking: The Science and Lore of the Kitchen.* New York: Scribner, 1984.

The Muscles in Our Ears

There are muscles in our ears that protect us from loud sounds, including the sound of our own voice. These muscles are in the middle ear. They're attached to the small bones connecting the eardrum to the cochlea, the chamber in the inner ear that contains sound-receptor cells. When these muscles contract, they dampen vibrations in the small bones, in effect muffling the sound before it reaches the inner ear. These muscles are not under voluntary control—they contract in a so-called acoustic reflex either just before a person speaks or just after a person hears a loud noise. Two specialists in hearing, Erik Borg and S. Allen Counter, wrote about the middle-ear muscles in the magazine *Scientific American*.

Apparently the contraction before speech has the function of protecting the speaker from the sound of his or her own voice. Borg and Counter estimate that without the acoustic reflex, the sound of a baby's crying would reach the baby's own ears with about the same intensity as the sound of a nearby passing train.

The contraction of the middle-ear muscles is especially effective at screening out the low-frequency components of the speaker's own voice, preventing high-frequency sounds from being drowned out. Many of the most important sounds in speech have high frequencies, so the acoustic reflex enables a person to understand the speech of others, even while speaking.

The middle-ear muscles also contract one or two tenths of a second after a loud external sound. That's fast enough to protect the inner ear from loud natural sounds such as thunder, but not fast enough to muffle a gunshot. Borg and Counter suggest that soldiers hum just before firing, to stimulate the acoustic reflex and protect their hearing. Another hearing expert once recommended that, for the same purpose, large guns be equipped with devices to generate a loud tone just before firing.

Borg, Erik, and S. Allen Counter. "The Middle-Ear Muscles." *Scientific American*, August 1989.

Rossing, Thomas D. *The Science of Sound.* 2nd ed. Reading, Mass.: Addison-Wesley, 1990.

Picket–Fence Echoes

Stand in front of a picket fence, clap your hands, and listen to the musical quality of the reverberation. The reverberation from a picket fence is made up of individual echoes of your handclap from each of the slats in the fence. The echo from each slat reaches your ears a tiny fraction of a second later than the echo from a neighboring slat a few inches closer to you. That's because the sound's round-trip travel time—from your hands, to a slat in the fence, and back to your ears—is shorter for slats nearer to you than for slats farther away.

If the slats are evenly spaced, as they usually are in picket fences, then the echoes reach your ears at regular intervals and you perceive the train of echoes as a musical tone. Notice that the pitch of this tone depends on the spacing between slats in the fence, not on the quality of your handclapping.

But we need to refine this story by pointing out that the musical tone you hear will not have a constant pitch. The tone will start out high in pitch and quickly sweep down to a low pitch. The high-pitched sounds come from the slats nearest you, which are all at just about the same distance from you. So echoes from those slats reach your ears very close together, giving you the impression of a high musical pitch.

The low-pitched part of the reverberation comes from the slats at the far ends of the fence. Each of those faraway slats is at a quite different distance from you than its neighbors. So echoes from those slats are separated by bigger time intervals, and you hear a lower pitch.

You might recognize this as being another example of the Doppler effect. Incidentally, you may hear the same "picket fence" effect by clapping near a long staircase or a corrugated wall.

Humphreys, W. J. *Physics of the Air.* New York: Dover Publications, 1964.

Heating Your Kitchen with the Refrigerator 175

Turn on your oven, and you'll warm up the kitchen. With the oven door open, the kitchen warms up even faster. That much is obvious since the purpose of an oven is to make things hot. But the opposite is not true of your refrigerator. Running the refrigerator makes the room warmer, and if you leave the door open, the kitchen warms up even faster. The first rush of cold air may cool things down a little, but in the long run the room will get warmer.

To see why, we need to think of heat as energy and cold as a lack of energy. The stove produces heat, but the refrigerator can't actually produce cold. All the refrigerator does is move heat, or energy, from one place to another. As the food inside the refrigerator loses its heat—or, in other words, gets colder—that heat ends up in the kitchen. Physicists call this kind of system a "heat pump."

But like any motor, the heat pump in your refrigerator needs energy just to run. So while it's busy moving energy out of the fridge and into the kitchen, it's also drawing in more energy in the form of electricity or gas. Since some of that energy is released as heat, you end up with more heat in the kitchen than you started with.

Air conditioners can cool your house because part of the unit is outside. That way the air conditioner can pump the heat out of your house and release it to the outdoors. So just as your refrigerator heats your kitchen while cooling the food, air conditioners heat the outdoors while cooling your house.

Listening Underwater

If you've ever been underwater at a pool when someone jumped in near you, you know that the sound of the splash is clearly audible. But telling where the splash came from is another matter. Even though water does a much better job than air of conducting sound waves, that extra conductivity makes it harder, not easier, to tell where a sound comes from.

Above the surface of the water, we can tell whether a sound comes from the left or the right because it strikes one ear a little sooner and a little more loudly. The more distant ear gets a smaller dose of the sound a little bit later because it's farther away from the source and also because it's shielded by the head. Even though we don't notice the difference consciously, it's enough for the brain to decide which direction the sound came from. But sound travels five times faster in water than it does in air. Traveling that fast, the sound is detected by both ears at almost exactly the same moment. That's one reason that underwater a sound seems to come from all directions at once.

The other reason is that underwater sound waves pass directly into your head, bypassing your ears altogether. That's because body tissues contain such a large amount of water. Try plugging your ears underwater and listening for another splash of someone jumping in. It will be just as loud as the last splash when your ears were not plugged. With sounds coming into every part of your head at almost exactly the same time, it's no wonder the brain has trouble deciding what direction the splash came from.

While you're in the pool, try this next demonstration with your friends. Duck your head underwater and listen to the conversation. If they talk loudly enough, you'll hear the vowels—*a, e, i, o,* and *u*—but no consonants. So the words won't make sense.

Sound travels very well underwater, but some sounds have more trouble than others getting from the air into the water.

But why the vowels and not the consonants? Every spoken sound is actually a combination of different sounds, some low, some high. Even though we don't notice the different sounds, the way they're combined is what gives each spoken sound its own character. In general, consonants contain a lot more high-pitch sounds than vowels. Those are sounds made of faster, smaller sound waves. Compared to consonants, vowels are mostly made of low pitches, in other words of larger, slower sound waves.

When the small sound waves hit the uneven surface of the water, they get scattered in all directions like ping-pong balls landing on a rough road. The much larger, lower-pitch waves aren't affected as much by the little water-waves because they hit a much wider area on the water's surface. If we think of a small sound wave as a little ping-pong ball on a rough surface, a larger sound wave is more like a big basketball, which is less affected by little bumps on the road. Unlike balls bouncing on a road, sound waves pass through the water, but like the basketball, the large waves come through with less distortion. That's one reason why when you listen underwater to someone up above, you won't hear the consonants with their high-pitched sounds and short sound waves.

Miller, Mary K. "Science in the Bathtub." *Exploring*, Winter 1993.
Schiffman, Harvey Richard. *Sensation and Perception: An Integrated Approach.* 2nd ed. New York: John Wiley and Sons, 1982.

A Major League Balancing Act

Try balancing a baseball bat straight up and down on the palm of your hand. When you let go with the other hand, the bat starts to fall. But if you're quick, you can move your hand in the same direction to keep it upright for a little longer. With some experimenting, you'll find that the bat stays up longer if you put the heavy end up. That's because in that position it falls a little more slowly, which gives you more time to respond before it falls off your hand.

But why does it fall more slowly with the heavy end up? After all, if you drop a light ball and a heavy ball, they should fall at about the same speed. The difference is that as the bat starts to fall, the bottom of the bat stays still on your hand while the top of the bat moves in an arc, both down and to the side. So gravity has to pull the bat down *and* get it moving horizontally at the same time. Even though a light object and a heavy object will *fall* at the same speed, it takes longer to get the heavy object moving horizontally. That's why pushing a VW is easier than pushing a Cadillac. Next time you're at the circus, watch the balancing acts. You'll see that whenever possible, the objects they're balancing will be weighted at the top.

Cross-eyed Cats

Have you ever noticed that Siamese cats are cross-eyed? In fact, that's the only way a Siamese cat can see straight. Unlike the eyes of some animals such as rabbits, a cat's eyes both point forward—just like ours—so most of what it sees it sees with both eyes. In order to see clearly, however, the brain has to coordinate the signals it gets from a group of nerve endings called the retina on the back of each eye.

For every spot on the retina of one eye, there's a spot on the retina of the other eye that has to see the same thing. Let's say, for example, that those spots are focused on a mouse. For the brain to interpret what it sees as one mouse instead of two, the nerves that detect the mouse in one eye have to go to the same part of the brain as the nerves in the other eye that detect the same mouse. If the eyes send the mouse images to two different parts of the brain, the cat sees two mice instead of one.

And that's what's wrong with the eyes of a Siamese cat. Instead of being lined up in the back of the eye, the center of the left retina is shifted to the right, and the center of the right retina is shifted to the left. So if a Siamese cat's eyes were pointed straight ahead, its retinas would be looking in different directions, sending a very confused message to the brain. By turning its eyes in, a Siamese cat looks cross-eyed, but its retinas are now lined up like a normal cat's, sending the brain a clearer picture.

Stent, G. C. "Explicit and Implicit Semantic Content of the Genetic Information." In *Foundational Problems in the Special Sciences,* ed. Robert E. Butts and Jaakko Hintikka. Dordrecht, Netherlands: Reidel Publishing Company, 1977.

Stent, G. C. "Strength and Weakness of the Genetic Approach to the Development of the Nervous System." *Annual Review of Neuroscience,* 1981.

Becoming Part of the Music

If you get close enough to the speakers at a loud concert, you can actually feel the low notes vibrating in your body. The higher notes may be just as loud, but you don't feel those.

Unlike waves on a lake, sound waves don't travel up and down. Instead they're more like layers of high and low pressure traveling outward in all directions. Each wave consists of a layer of high pressure followed by a layer of low pressure. The bigger the difference between the high pressure and the low pressure, the louder the sound.

But sound waves don't just travel in air: except for a vacuum, they travel in whatever they encounter, including your body. So when you listen to loud music, your whole body alternates between high and low pressure just like the air around it.

The reason high and low pitches feel different has to do with the length—or thickness—of the wave. High-pitch sounds produce thousands of very short waves—maybe 8 to 10 inches thick—each second. A very low note might generate only about 65 waves per second, but they could be nearly 17 feet thick.

That difference affects the way you *feel* the music because with a low note, your body spends a relatively long time actually inside the layers of low and high pressure. The waves of a high note travel through your body as well, but they oscillate so quickly between high pressure and low pressure that on the average you don't feel any real pressure change. So next time you're at a concert and you feel those low notes going through you, remember that you are a *part* of the music.

179

Why It's Hard to Burn One Log

Imagine a cheerful fire, with several wood logs piled up in the fireplace. Separate one of those logs from the others, and it will usually stop burning. Separate all the logs from each other, and the flames soon die away.

Scientists who have studied the burning of wood in detail have found why it's hard to keep one log burning. The flames that envelop a burning log are fed not so much by the solid wood near the surface but mostly by flammable gases coming from inside the log. Those fuel gases, in turn, are released when molecules making up the solid wood are broken down by heat from the fire. A log won't burn unless layers of wood below its surface are kept hot enough to continue producing those flammable gases.

One log, burning by itself, cannot generally send enough heat into its own interior to continue producing those gases. Most of the heat from a burning log travels away from the log, either in the form of hot gases and glowing soot particles, or as infrared light which is sometimes called radiant heat energy. So one way to keep a single log burning would be to put a concave metal mirror near it. The mirror would provide a reflecting surface to return some of the heat that would otherwise escape from the log.

Of course, a mirror near a burning log is likely to be blackened by soot. A more practical way to keep one log burning is to put another burning log next to it. Then each log captures radiant heat energy from its neighbor. The interior of each log is heated not only by the flames on its own surface but by heat captured from the neighboring log. That extra heat keeps the interior of each log hot enough to continue producing fuel gases, and the fire can continue to burn.

Lyons, John W. *Fire.* New York: Scientific American Library, 1985.

Mushrooms and Rocket Fuel

Perhaps the most popular wild edible mushrooms are a group called the morels. A similar group, known as false morels, also contains some edible species, but just how edible false morels really are is something of a tricky question.

What once puzzled both scientists and mushroom hunters was the way people reacted to these mushrooms. According to popular wisdom, the cook was much more likely to get sick than the guests. And when the guests did get sick, it was usually only one or two, while the rest felt fine. One theory held that it was simply an allergic reaction.

The explanation came with the development of the space program, when someone noticed that workers exposed to rocket fuel showed the same symptoms as people poisoned by false morels. The culprit turned out to be a group of chemicals called hydrazines, which are both a major component of rocket fuel *and* one of the poisons found in false morels.

The way hydrazines affect the body, there's a fine line between a poisonous dose and a harmless dose. So if one person eats just a little more than someone else, he or she could become terribly sick while the other wouldn't feel a thing. Also, hydrazines evaporate very easily. So cooked mushrooms contain much less toxin than raw mushrooms. And the cook, who breathes the vapors, is more likely to get sick than the guests who eat the mushrooms after the hydrazines have been boiled out.

Other mushrooms, including the ones in the grocery store, contain hydrazines, but in even smaller amounts than the false morels. And if you cook your mushrooms, even those small amounts will evaporate.

Ames, Bruce, and Lois Swirsky Gold. "Dietary Pesticides (99.99% All Natural)." *Proceedings of the National Academy of Sciences* 87 (October 1990).

Lincoff, Gary. *The Audubon Society Field Guide to the North American Mushrooms.* New York: Alfred A. Knopf, 1981.

Lincoff, Gary, and D. H. Mitchel. *Toxic and Hallucinogenic Mushroom Poisoning: A Handbook for Physicians and Mushroom Hunters.* New York: Van Nostrand Reinhold Company, 1977.

Lest you get the impression that all science is based on incontrovertible facts resulting from extensive investigation, these next three pieces should let you know that sometimes more research is necessary. And since some of this research is unique, we are always careful to air reports on it on the appropriate day in the spring of the year. After all, timing is everything.

Where Do Bulbs Go in the Winter?

If you actually enjoy planting flowers, you probably prefer flowers such as pansies, nasturtiums, and marigolds that have to be planted every year. If you like flowers but prefer to keep your knees out of the mud, you may find bulbs more to your liking. Plant a batch of daffodils, crocuses, or day lilies, and they come back year after year.

But where do the bulbs go when they're not in bloom? Every spring the crocuses appear, followed by daffodils, then tulips and hyacinths. But soon each one disappears until the following year. Older theories speculated that the bulbs simply stayed put, but that doesn't make sense since they wouldn't have anything to do and would get so bored they'd probably die.

In fact, the truth wasn't revealed until 1979, when a major earthquake opened the ground in Central America to reveal thousands of migrating hyacinth bulbs. Since then, botanists have determined that between nine and eleven days after pulling in their leaves, bulbs gather in large flocks. Huddled together for protection against predators, the bulbs begin a long migration to their communal spawning grounds in the South American rain forests.

As spring arrives, newly hatched bulbs leave their dying parents behind and head for the northern climates. Crocus bulbs, being the smallest, travel fastest through the ground, and that's why they appear first.

How bulbs navigate the long and arduous journey to a destination many have never seen is still a source of wonder. According to one theory, they communicate telepathically with high-flying ducks. Of course there's no good reason for this theory, but if it's scientific, who needs reasons?

A Taste for the Musical

Have you ever noticed that you, your children, and your parents all like different music? Of course you have, but you may not realize the biological basis for these differences.

As it turns out, a patch of taste buds located on the inner ear is controlled throughout an individual's life by various chemicals released in the pituitary gland. During puberty, the pituitary gland releases large quantities of an enzyme known as Zepellinase. As Zepellinase is released by the pituitary gland, it travels to the inner ear, where it binds to heavy metal receptors located on the cell walls. Zepellinase somewhat blocks the effectiveness of the taste buds, causing a need for excessive volume and preventing the reception of more than one or two chords during any given piece.

As the body's chemical systems settle into a middle-aged routine, Zepellinase is replaced by the hormone Vivalderone, causing the inner ear to crave large doses of baroque fugues. A debilitating though lamentably common hereditary condition causes the pituitary gland to release an excess of Vivalderone, producing a craving for harpsichord inventions—the more parts the better.

Sometime into senescence, the taste buds fall off altogether, at which time the ear loses all basis for filtering, screening, or otherwise discriminating. A frequent result of this late condition is a desire for the recordings of Lawrence Welk and Barry Manilow. Fortunately, however, a group of scientists led by Dr. Karl Haas are already investigating the use of dietary supplements as a means of reversing these traumatic effects.

Where Do the Socks Go?

This Moment of Science is going to deal with one of the most perplexing mysteries to confront modern societies: the disappearance of socks in a clothes dryer. It doesn't seem possible that you can put five pairs of socks in the dryer, and have only four and a half pairs left at the end of the drying cycle. Scientists do not yet fully understand this phenomenon, but here is a report on a recent promising theory.

As the socks tumble in the dryer, water molecules are shaken from them into the warm air in the dryer. An exhaust fan draws the warm, moist air to the outside. Eventually all the water molecules leave the socks and are exhausted to the outside.

Scientists now theorize that on occasion there are chemical changes that take place which affect a whole sock, not just the water molecules. There is evidence that in about every fourth to sixth load, the agitated environment in the dryer creates an atmosphere with abnormal concentrations of dimethyl terphthalate, ethylene glycol, and ozone. The first two chemicals are essential raw materials in the manufacture of polyester fibers, and ozone is, well, the stuff with the big hole in it. The compound that results has been named hozone.

For reasons not fully understood, when the dryer has such a high concentration of hozone, one sock in the load is converted into these component chemicals and shunted off into an as yet undiscovered area called the hozone layer.

To date this is the most plausible explanation for the missing socks, but it has also raised another concern: Scientists are investigating the possibility that the heat-absorbing qualities of the billions of socks in the hozone layer might be having some effect on the environment, although they don't yet know if the mysterious layer is contributing to global warming or cooling.

Index

187

DON GLASS has worked in public radio for thirty years. He has a number of national productions to his credit, and has been associated with the radio series *A Moment of Science* since its inception in 1988. His producing and writing of the radio programs has allowed him to combine his professional radio career with his lifelong interest in science. He also edited this book's predecessor, *Why You Can Never Get to the End of the Rainbow, and Other Moments of Science.*

Stephen Fentress is on the staff of the Strasenburgh Planetarium of the Rochester Museum and Science Center in Rochester, New York, where he writes and produces planetarium shows. *Barbara Bolz, Eric Sonstroem,* and *Don Ulin* are doctoral candidates in the English Department at Indiana University. Their writing skills have allowed them to share with a large audience their fascination with the world around them, first through the radio programs and now through this book.